Phosphate Coatings Suitable for Personal Protective Equipment

Diana Petronela BURDUHOS-NERGIS

Costica BEJINARIU

Andrei Victor SANDU

Published by **Materials Research Forum LLC**
Millersville, PA 17551, USA

Published as part of the book series
Materials Research Foundations
Volume 89 (2021)
ISSN 2471-8890 (Print)
ISSN 2471-8904 (Online)

Print ISBN 978-1-64490-110-6
ePDF ISBN 978-1-64490-111-3

Distributed worldwide by

Materials Research Forum LLC
105 Springdale Lane
Millersville, PA 17551
USA
http://www.mrforum.com

Printed in the United States of America
10 9 8 7 6 5 4 3 2 1

Table of Contents

Introduction

Out of the need to be safe during escalations, the first safety rings were invented in the early twentieth century, and then in 1910, Hans Fiechtl invented and manufactured the python. The geographical extension of mountaineering was an important factor in modernizing the altitude protection, the python became a mandatory accessory, although it was difficult to use. This problem was solved with the appearance of the carabiners.

A carabiner is a special type of metal link with a spring-loaded locking element, used to quickly and reversibly connect components, mainly to safety systems.

The choice of material is a main step in the carabiner design process, the materials being chosen according to the areas in which the carabiners are used and the properties they must have. Carabiners made of aluminum alloys are used in places where the specific weight plays an important role, for example, in sport mountaineering. While, steel carabiners are used in places where high mechanical strength and refractory properties are required, for example, in rescue operations during fires.

The carabiners are currently used in many fields, such as the oil industry, speleology, construction, arboriculture, etc., so there are a number of main factors that must be taken into account when designing them. One of the factors that can affect the properties of the carabiners is the temperature. They must withstand large temperature differences, for example, from negative temperatures (-40 °C) during mountain climbing to high temperatures (+500 °C) when used by firefighters during rescue, evacuation and extinguishing fires. The coefficient of friction is also an important factor, as a rough surface can lead to premature wear of the ropes. Due to the places where they are used, carabiniers come into contact with various corrosive substances, such as salt water or fire extinguishing solution, so one of the most important factors to consider is corrosion resistance. Because carabiners are used at high altitudes, they can be hit or dropped on various hard bodies, diminishing their resistance due to the appearance of cracks or internal microcracks.

According to one of the largest carabiner companies, one of the most widely used materials in the manufacture of carabiners is carbon steel. It has both good mechanical strength and refractory properties. Due to the use of carabiners in environments where they come into contact with salt water or various corrosive substances, as well as the possibility of being hit by various bodies, these connecting elements must possess good properties of corrosion resistance and shock resistance.

Over time, the carabiniers have undergone a series of changes both in design and in shape or materials. However, there are still many challenges to be overcome in protecting against mechanical shocks and corrosion of the materials from which their bodies are made. Following the critical analysis of the literature on the failure of personal protective equipment against falling from a height, due to damage to the connecting elements, it was found that the durability of carabiniers is affected by several factors that take into account the load conditions, the environment and their geometry. Thus, it was observed that the decommissioning of carabiners made of carbon steel is due to the appearance of iron oxides on their surface or as a result of suspicions regarding the appearance of internal cracks. Therefore, the book addresses the improvement of the corrosion resistance of carbon steel used in the carabiners manufacturing, as well as the development of a method to protect them against mechanical shocks by depositing layers based on elastomers.

Improving the corrosion resistance of carbon steel carabiners can be achieved by depositing a layer of insoluble phosphates on the surface of the material. The specific porosity of the phosphate layer allows the deposition of an elastomer-based paint that absorbs mechanical shocks that can lead to cracks in the material from which the carabiner body is made.

The book is relevant for fundamental and applied research in the field of materials engineering because it shows the obtaining of phosphate layers with industrial applications. It also describes how to design and develop phosphating solutions that differ in the type and concentration of metal ions dissolved in phosphoric acid. Moreover, the deposited layers are characterized, by specific analyzes in laboratory conditions, from a chemical, structural and mechanical point of view.

The first chapter, "Personal Protective Equipment Against Falls from Heights and its Components"; presents introductory notions related to personal protective equipment for fall protection and their components.

In the second chapter, "Materials used to Manufacture Carabiners" are presented and characterized the materials from which the carabiners are made, their manufacturing process and the properties they must have according to the international standards. At the same time, it contains a critical analysis on the physical, mechanical and chemical properties of the main materials used in the carabiners manufacture.

The third chapter, "Objectives and Methodology of Experimental Research", presents the research focus, the material on which the phosphate layers were deposited and the methodology and technology for obtaining the phosphate layers. Further, are presented the characterization methods of the obtained deposited layers from: (i) elementary and structural point of view: by X-ray spectroscopy by energy dispersion, optical microscopy, scanning electron microscopy and by Fourier transform infrared spectroscopy, (ii) mechanical point of view by scratch, microindentation and impact tests, (iii) chemical point of view by corrosion resistance methods. At the same time, this chapter briefly describes the techniques and the equipment used to analyze the obtained layers.

The fourth chapter, "Structural Characterization of Phosphate Coats" contains experimental results obtained from the evaluation of the chemical and structural characteristics of the three types of phosphate layers deposited by conversion on the carbon steel surface.

The fifth chapter, "Mechanical Characterization of Coats", presents the mechanical characteristics of the deposited layers determined by scratch and microidentation tests, as well as by impact test.

The sixth chapter, "Corrosion Resistance of the Deposited Layers", the corrosion resistance of the deposited layers and the base material are analyzed and compared. The corrosion resistance evaluation was performed by linear and cyclic polarization, as well as by electrochemical impedance spectroscopy, in three corrosive environments: rainwater, seawater and fire extinguishing solution.

The last chapter entitled "Design and Applicability" presents the successful application of the deposited phosphate layers and indicates possible future directions of research in this field.

CHAPTER 1

Personal Protective Equipment Against Falls from a Height and its Components

1.1 Personal protective equipment

Personal protective equipment, known as 'PPE', is equipment used to minimize exposure to risks that could cause serious injury at work and occupational diseases. These injuries and diseases may occur as a result of the worker's exposure to chemical, radiological, physical, electrical, mechanical or other risks. Personal protective equipment may include gloves, noise pollution devices, goggles, footwear, helmets, respirators, overalls, harnesses, etc.

All personal protective equipment must be designed and manufactured to provide safety in accordance with specific standards. In order to encourage their use by the worker and not to expose them to other risks, PPE must be comfortable, match the size of the user and cover the risk for which it is used. Where risks cannot be avoided or mitigated by intrinsic or collective protection measures, employers must provide personal protective equipment and ensure that it is used properly. The necessary personal protective equipment is selected according to the risks to which workers are exposed, determined by work accident and occupational disease risk assessment [1,2].

Personal protective equipment that is used against falls from a height is required when the worker is at risk of falling while working at a height. The main causes of accidents due to falls from a height are the lack of appropriate personal protective equipment, the use of worn or torn equipment or its improper use [3].

'Personal protective equipment against falls from a height' means all elements and components fitted in a certain order in order to prevent the worker from falling from a height and suffer injuries.

According to statistics, the consequences of a fall can be dramatic, regardless of the height where the workers carry out their duties [4]. The most common accidents are falls from a height, and they can have serious consequences such as severe injuries, bone fractures, rupture of internal organs, massive internal bleeding or even death.

Falling from a height may also be caused by equipment touching abrasive materials or sharp edges, as well as by environmental factors such as humidity, high or low temperature, etc. Personal protective equipment must preserve its characteristics in all foreseeable environmental conditions for a sufficiently long period of time and any change affecting its performance must be easily noticeable [5].

Personal protective equipment against falls from a height is a set of components designed to protect the user and consists of:

-      a safety system including an anti-fall harness, a positioning belt during the activity, an energy absorber and a rescue lifting device;

-      an attachment system, which includes an anchoring device that can be connected to a stable attachment point (carabiners or hooks).

Here are the main types of safety systems against falls from a height and their roles:

-      restraint systems, which prevent the user from reaching an area where there is a risk of falling from a height;

-      positioning systems during an activity, which prevent free fall, while allowing the user to work at a height, suspended;

-      cable access systems, which allow the user to access to and from the workplace in such a way as to prevent or block their free fall by using a working line and a safety line connected separately to the reliable anchor points used for fixing the workplace or for rescue;

-      fall arrest systems that reduce the impact force exerted on the user's body;

-      rescue systems that help the user to prevent other people or themselves from a free fall.

The fall arrest system is personal protective equipment that prevents the user from colliding with the ground or any other obstacle during a free fall and limits the force of impact on the user's body when coming to a stop [6].

European directives on work health and safety, including temporary work at a height, such as Directive 89/655/EEC as amended by Directive 2001/45/EC, require that, following risk assessment, employers take technical measures to prevent falling from a height while performing work tasks. The directives give priority to collective protection measures, such as: solid swings, scaffolding and platforms or safety nets. The special provisions of this Regulation are intended to use cable access and positioning techniques. Unlike other older national laws requiring special protection measures for heights of more than 2m, the European Directive 2001/45/EC [5] does not set a limit for the minimum distance from the ground from which special protective measures must be taken [8].

The European Directive on the use of personal protective equipment (PPE) at work lays down the main rules to be followed by the employer to ensure that personal protective equipment is appropriate and will protect the user/worker [9]. This directive recommends the use of harnesses (for scaffolding work, assembly of prefabricated parts, masonry work) or cables (for work in high crane cabins, work in high forklift cabins, work in high sections of drilling rigs, work in wells and canals).

The components of personal fall protection systems are personal protective equipment included in Directive 89/686/EEC, as amended by Directives 93/68/EEC, 93/95/EEC and 96/58/EC, which specify the basic requirements for the health and safety of workers, and other criteria to be met by personal protective equipment sold on the internal market or used in the European Union. The manufacturer is fully responsible for ensuring PPE compliance with all the legal provisions in force.

Any fall arrest system should include:

- a suitable body support device, i.e. a complete harness;

- an energy absorbing element: energy absorber, retractable drop stop, guided drop stop;

- a fall protection guide and a flexible support;

- an anchoring line;

- an anchor point;

- connectors (carabiners or hooks).

Most manufacturers deliver PPE components against falls from a height that can be assembled into different fall arrest systems. Manufacturers offer recommendations on the compatibility of products with other elements, so that the employer can assemble them to obtain a suitable anti-fall system. In both cases, the responsibility of checking whether the system is appropriate to the activity to be performed and the working conditions lies with the employer [10].

When combining components in order to build a fall arrest system, the following aspects must be taken into account:

- adequacy of components for the intended use of the fall arrest system, taking into account all the different stages of use (e.g. access to work);

- workplace characteristics (e.g. floor tilt, anchorage location, need for free movement over long distances, other places in the work space where a fall could occur).

The actual working height of the worker must be greater than or at least equal to the minimum distance indicated by the manufacturer of the fall arrest;

- targeted user (for instance, level of competence, experience);

- compatibility of components (for instance, interaction between anchor and other components);

- ergonomic considerations (for instance, right choice of ropes and attachments to minimize discomfort to the body);

- information provided by the manufacturers for all components;

- the need to facilitate safe and efficient rescue operations (for example, to prevent injuries caused by suspension at a height, taking into account that a healthy person suspended upside down loses consciousness within $6 \div 20$ minutes after a fall);

- anchorage characteristics (for example, location, strength and shape).

Experts believe that there are ten criteria that each worker must apply to individual fall protection systems when choosing the equipment needed to perform their tasks safely and in accordance with the applicable law. This involves, in addition to the above criteria, the traceability and monitoring of the degree of wear of the fall arrest system, taking into account previous exposure to shocks, aggressive agents and natural aging [11].

The components of fall protection systems used while working at a height are: anchoring devices (this equipment is intended to provide an anchorage from which a safety rope or retractable fall arrest can be attached), positioning belts (these belts are part of the category of safety belts that aim to position the worker during work and must be used as a means of supporting the body), complex belts (the main functions of which are the positioning of the worker during work, limiting the movement of the worker towards the source of injury by falling from a height and positioning and suspending the worker while working), fall arresters (devices that can be mounted on flexible anchor brackets or on rigid elements positioned above the work area), connectors (carabiners or hooks). Such a system is shown in Fig. 1.1.

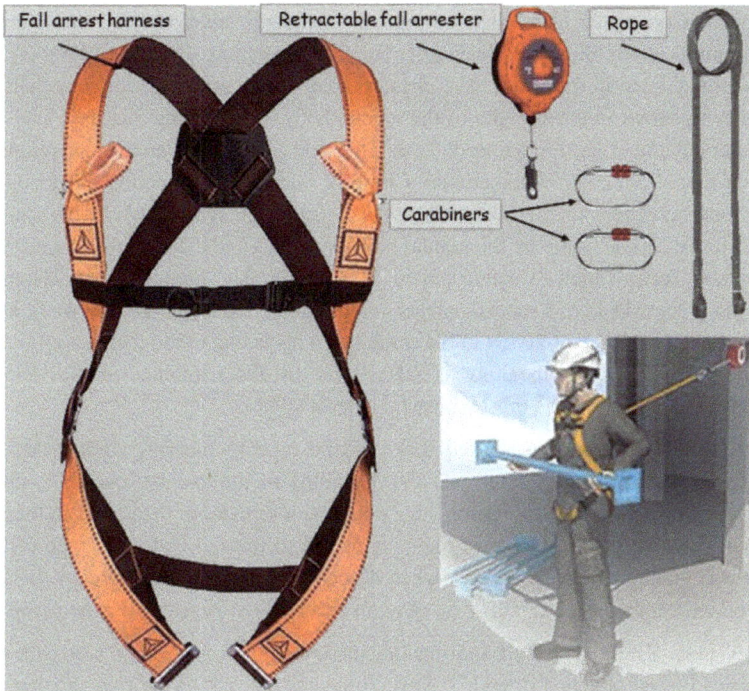

Figure 1.1 Fall protection system [12].

1.2 Carabiners

A carabiner is a special type of metal hook with a spring-loaded locking element [13] used to quickly and reversibly connect components, mainly to safety systems [14]. The word is an abbreviated form of the word Karabinerhaken, a German expression meaning 'spring hook' [15] used by carabineers to attach items to a belt or shoulder strap.

Carabiners have been improved several times over the decades, adjusting their mass, reliability and durability.

1.2.1 History

European standards are implemented by one of the three European standardization organizations (ESOs): CEN (European Committee for Standardization), CENELEC

(European Committee for Electrotechnical Standardization) or ETSI (European Telecommunications Standards Institute). These organizations conducted tests on ropes to standardize them in the 1960s. The physical basis for all standards for ropes and other related components is the strength of the human body. Considering that the value of the Earth's gravitational field is 1 G or 9.8 m/s^2, the safe fall stop distance ranges from 1/10 (physical) to 1/5 (UIAA-International Climbing and Mountaineering Federation) of active rope according to the UIAA/CE rope dynamics standard [16]. In this case, the generated force to be borne by the human body is about 12 kN. This value comes from a US military specification implemented in parachuting. The harnesses are designed to withstand at least 15 kN. A second category fall with an impact on the user of 12 kN can easily propagate a force of 20 kN in the carabiner. This is due to the fact that up to 1/3 of the force discharged on the knot is taken over by the carabiner. It was therefore concluded that all carabiners must withstand at least 10 kN.

Before carabiners, a cord was used that tied the rope to the ring of a metal piton, providing a fixed means of safety. At the beginning of the twentieth century, the first (safety) rings were designed in Europe that could be inserted into the cracks in the rocks. They consisted of iron spikes with rings connected to the flat end. This type of safety device was mainly designed to serve when descending, as additional support for feet or hands on the climbers' route. This type of piton was made to be used, in general, by hand.

In 1910, Hans Fiechtl invented and manufactured the piton made of a soft part with a hole instead of an attached ring [17]. This eliminates the risk of the ring sagging under stress. There were no official standards for these parts of the equipment because the manufacturing methods had huge inconsistencies. Nowadays, the UIAA 122 standard sets the hardness for hard and soft pitons, as well as the maximum load limits in the three planes for this type of safety device.

The geographical extent of mountaineering has also been a determining factor in the technological process. The year 1927 brought two changes on protection, one in America and another, more advanced, in Europe. Joe and Paul Stettner ordered pitons with twisted rings from Munich, went to Colorado, bought a rope from a local store, and climbed the east face of Long Peak [18]. This climbing marked the first example of mechanically protected mountaineering in America. Lead and then soft steel pitons in Europe and Colorado were not suitable for the granite mountains of Yosemite National Park. In 1946, the climber and blacksmith John Salathe used ultra-strong pitons made of high-carbon steel, chromium and vanadium, which could be inserted into granite without being destroyed or flaming [19].

In Europe, pitons were already mandatory for all climbers, with the development of more advanced techniques for their protection. The same year that Stettner climbed the Long Peak, climber Laurent Grivel invented and manufactured the rock drill and the expanding screw [17]. These items would become famous, because they are the primary means of adding extra permanent protection on the climbing routes around the world even today. Despite their revolutionary aspects, they have not been part of the equipment of climbers for many generations.

The difficulty of pitons was overcome by the invention of carabiners. This was the name used by German and Italian soldiers for the connector that attached the rifle to their belt around the 1900s.

In 1919, the first efforts were made to standardize the equipment of climbers in the 'Guido Rey' paper published in the Aplinisme Acrobatique journal [20]. After the First World War, the international mountaineering organization easily scattered this information in the climbing areas. They standardized the techniques by introducing pitons and carabiners. The post-war period also saw an improvement of the quality of rope fabric and carbon-steel carabiners [17].

In 1910, Rambo Herzog came up with a new idea for climbing carabiners, when he saw a brigade of firefighters wearing elongated staples to their belts. He implemented this idea in the world's first climbing carabiner, although previous versions of it were unsafe, as carabiner gates could be easily opened with one hand, but could not be guaranteed to remain closed throughout the climb [21].

In 1938, two world famous climbers, Paul Allain and Raffi Bedayn, both developed much lighter and more reliable aluminum carabiners. While some sources report that Paul Allain came up with the idea first, Bedayn carabiners were the ones to gain fame and notoriety. This may also be due to the fact that Raffi Bedayn belonged, at that time, to one of the largest sports climbing clubs, the Sierra club [17].

In any event, the concept behind the carabiner has hardly varied since its invention, in terms of basic style and shape. They are still based on 'coupling links with a safety locking element'. Pear-shaped carabiners are the most common on the market, followed by oval ones and then by those with more complex shapes [22]. However, there were also some innovations in terms of the design, material or shape of the locking element that occurred in various places depending on the targeted users. All climbing carabiners are now made of solid metal. Yet, in the 1970s, Salewa introduced a model made of pipe that weighed only 40 grams. This model was not only revolutionary because of its shape, but also because of the safety tests performed on each unit [23]. For the first time, each carabiner was tested individually before being placed on the market. The little groove on the curvature of the carabiner is the sign of the 1000 kg test. Many climbers looked at the hollow interior and imagined that these empty carabiners were not safe. However, in an interview with the 'Alpinist', the former general manager of Salewa, Hermann Huber, claimed that the hollow models were abandoned due to cracks during cold forging, and they were replaced by solid aluminum that were much lighter and durable.

The first locking elements of carabiners, made of wire, were the work of an experienced navigator, but Hotwires Black Diamond was not influenced by the Kong's 'sea gates'. Andrew McLean spoke of the wire safety elements and staples that he used when sailing. McLean told the 'Alpinist': 'I did not work for Black Diamond for a long time and I did not realize how deeply-rooted the standard locking element was. Therefore, I came up with a carabiner prototype the locking element of which was made of wire'. The company was reluctant about this type of locking element. McLean's colleagues feared that this type of locking element would not be accepted by the clients or that the new design would fail after just one year. This was not a surprising reaction, as Black Diamond had recently emerged after the bankruptcy of Chouinard Equipment Ltd. McLean credited Johnny Woodward to promote the project and did much of the initial testing [17]. One test included taking high-speed photos of carabiners while they were hit by a solid object. The traditional locking elements would have opened during this process, but the low-mass wire locking elements remained closed. The new design was not only easier to manufacture and smaller, but also safer.

Thus, carabiners became a much more versatile device, especially after being processed by the minds and hands of very intelligent craftsmen. There are now carabiners made of all kinds of combinations of materials, sizes, shapes, colors and there are also multifunctional carabiners that can be used to open bottles or as screwdrivers. From their humble beginnings as a simple war staple to the most important innovations in mountaineering (along with ropes, of course), to a multifunctional tool, carabiners have had a unique path.

1.2.2   Areas of use. Classification

Carabiners are widely used in activities such as: leisure mountaineering (Fig. 1.2a), utility alpinism (Fig. 1.2b), navigation (Fig. 1.2c), arboriculture (Fig. 1.2d), rescue operations (Fig. 1.2e), hot air balloon flying (Fig. 1.2f), in construction works (Fig. 1.2g), in acrobatics (Fig. 1.2h), speleology (Fig. 1.2i), as well as in the industrial field (Fig. 1.2j).

a. leisure mountaineering [24]

b. utility alpinism [25]

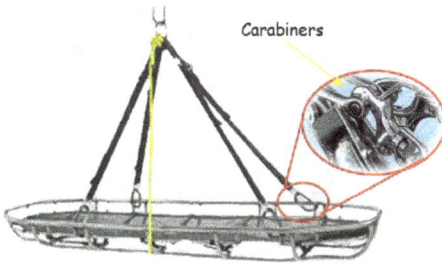

c. navigation [26]

d. arboriculture [27]

e. rescue operations [28]

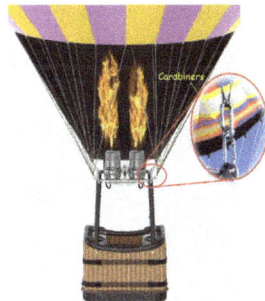

f. hot air balloon flying [29]

Materials Research Forum LLC
https://doi.org/10.21741/9781644901113

g. construction works [30]

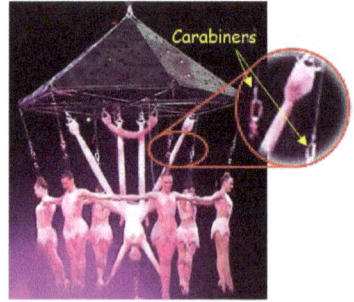

h. acrobatics [31]

i. speology [32]

j. oil industry [33]

Figure 1.2. Carabiners areas of use.

Carabiners used in sports tend to weigh less than those used in commercial applications or rescue operations. While from an etymological point of view, any metal hook with a locking element actuated by a spring is, technically speaking, a carabiner, in these areas mentioned above, only carabiners that were manufactured and tested for work systems at a height are used [34]. For instance, threaded carabiners are used to connect the diver's rope to the surface wharf wiring. These are usually rated for a safe workload of 5 kN or more (equivalent to a mass greater than approximately 500 kg) [34, 35]. According to the EN 12275:2013 standard, the components of a carabiner are shown in Fig. 3.

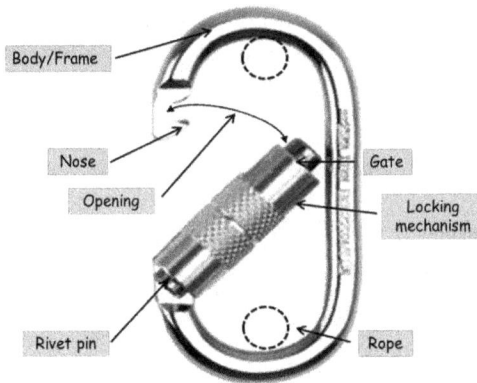

Figure 1.2 The carabineer components [14].

There are now many different shapes of carabiners that are used in all fields. Nevertheless, there are three main shapes: D-shaped, pear-shaped (HMS) and oval carabiners.

D-shaped carabiners

The D shape directs the load on the load support bar (main axis), which is the strongest point of the carabiner. This is a great advantage when it comes to quick coupling or uncoupling. Due to its design, the rope slides on the body of the carabiner, so the rope is kept away from the locking element, which minimizes the risk of unintentional decoupling [36].

HMS carabiners or pear-shaped carabiners

HMS carabiners have two main features: they have a very large opening of the locking element and a body with space for several knots. This shape prevents crossing and also reduces the risk of the knot moving and opening the locking mechanism of the locking element. Virtually all HMS carabiners are carabiners with locking systems [37].

Oval-shaped carabiners

Oval-shaped carabiners are common and highly used. Due to its symmetrical design, the load is always in the middle of the rear bar (main axle), thus allowing easy repositioning of loads. Carabiners are used in a variety of situations. This is why there are so many different types available. New models are introduced on the market every year. There are three main types of carabiners: non-load-bearing carabiners (accessory), basic or normal carabinieri (often referred to as 'non-locking' carabiners) and carabiners with locking system [38]. Table 1.1. shows the main advantages and disadvantages of the different types of locking elements, as well as the locking mechanisms to be taken into account when choosing a carabiner [39].

Table 1.1   Advantages and disadvantages of different types of gates and locking mechanisms.

Types of gates and locking mechanisms	Advantages	Disadvantages
With manual closing and manual locking	Cannot be opened accidentally. Superior resistance to all three load modes.	If the user forgets to block them, their resistance decreases exponentially. Handling time is longer.
With automatic closing and manual locking	The possibility of accidental opening is very small.	The user must remember to lock the gate.
With automatic closing and locking with spring sleeve	The user does not have to remember to lock the gate.	The possibility of gate accidental opening is very high.
With automatic closing and locking with spring sleeve and additional protection mechanism	The user does not have to remember to lock the gate. The possibility of accidental opening is very small.	Heavy handling of the gate.

Basic carabiners

Regular carabiners are basic connectors with a locking element, but without an automatic locking mechanism. They can be purchased either separately or in sets. Regular carabiners (basic carabinieri) are easier to handle than a carabiner with a locking mechanism, but can accidentally open much faster. When the ease and speed of coupling and decoupling are more important than the additional security provided by a locking mechanism of the locking element, a regular carabiner is used [40]. The locking element of this carabiner can be of two types:

a)   Bar-like locking element;

b)   Wire-like locking element.

Carabiners with locking system

These carabiners are classified depending on their locking mechanisms. Carabiners with locking system are always used to secure descents from glaciers or buildings. Carabiners with less complex locking mechanisms are used in rappelling to provide additional protection.

Although carabiners with a locking system provide higher safety, they must remain easy to handle. They are classified into two main groups: carabiners with manual locking system and self-locking carabiners. Carabiners with a manual locking system consist of a mechanism that must be closed by the user, while self-locking carabiners lock automatically as soon as the locking element is released [41].

a)   Carabiners with manual locking mechanism

These carabiners must be closed manually by the user. When it comes to handling, they require more closing time than self-locking carabiners. However, carabiners with manual locking mechanism have their own advantages. For instance, in a safe position, if the locking mechanism is not closed, the carabiner can be used as a regular carabiner. Depending on the locking mechanism used, there are two types of carabiners:

With threaded locking element

To lock the carabiner, screw in the sleeve after closing the locking element.

With double safety

These carabiners have a device that provides additional safety to the locking mechanism of threaded carabiners.

b)   Self-locking carabiners

Self-locking carabiners have a spring mechanism, which closes automatically as soon as it is released.

With twist lock

These carabiners have a twist-operated sleeve. Unlocking them is done by turning the sleeve by about 90°.

With sliding lock

EDELRID has developed a carabiner locking system that is easy to use with one hand. To open, the small bar on the outside of the locking element slides.

With three locks

Carabiners with three locks require three separate actions to open the carabiner. There are two different types: 'push and twist' and 'pull and twist'.

With ball lock

It is a special type of connection element with three locks: to open the carabiner, the ball is pressed and then twisted, which unlocks the ball locking cuff.

With sliding lock and safety system

These carabiners combine a sliding locking mechanism with an internal spring bar. As such, they have two fully automatic locking mechanisms that are completely independent of each other. The carabiner will only open when the spring bar is pushed up and the sliding mechanism is released [42].

In addition to the two main classifications related to the way carabiners are designed, the EN 12275:2013 standard ranks them according to their use, namely:

a)   Type B carabiner (basic) with sufficient strength to be used in a fall arrest system;

b)   Type H carabiner (marked HMS) intended for semi-cabestan knot securing ('Halbmastwurfsicherheit');

c)   Type K carabiner (kletersteig) which is intended for intermediate securing of the rope end during a climb;

d)   Type D carabiner (directional) which is intended for rope anchoring or guiding systems;

e)   Type A carabiner (Haken) which is intended for self-insurance on piton, anchor, etc.;

f)   Type Q carabiner (Quicklink) which has a fast locking system, by screwing;

g)   Type X carabiner (oval) is the type of carabiner used for small loads, like attaching the rappel cord to the anchoring point.

Their schematic representation, as well as the static strength on the longitudinal axis with the locking element closed (Rl_1), the static strength on the longitudinal axis with the locking element open (Rl_2) and the static strength on the transverse axis (Rt) are shown in Table 1.2 [35].

These types of carabiners have different sizes or loading gauges by design. For instance, type K carabiners require that the space between the locking element and the body should allow the insertion of a $\Phi21$ mm cylinder; for type B, H and X carabiners, this space must allow the insertion of two $\Phi11$ mm ropes.

Table 1.2   Carabiners classification according to the SR EN 12275: 2013 standard.

Type	Name	Scheme	Rl₁	Rl₂	Rt
B	Basic carabiner		20KN	7KN	7KN
H	HMS carabiner		20KN	6KN	7KN
K	Klettersteig carabiner		25 KN	8KN	7KN
A	Special piton carabiner		20KN	7KN	
D	With secured position		20KN	7KN	
Q	With threaded gate		25KN		10KN
X	Oval carabiner		18KN	5KN	7KN

1.2.3 Risks and hazards

Even the best carabiners can be a potential source of risk if not used properly. The main risks associated with carabiners and the dangerous situations to avoid are listed below.

The effect of opening the locking element by inertia

This effect of inertia occurs when the back of a carabiner, i.e. the side opposite the locking element, hits hard a solid surface (e.g. a rock). In the event of a fall, as the rope is pulled, the carabiner often hits against the wall. Due to inertia, the locking element opens for a brief period of time, and at that moment the risk of the operator falling freely increases considerably.

If load is applied on a carabiner while it is opened, then its breaking strength decreases by half, compared to the existing strength when the locking element is in the closed position. The forces generated during a fall can, in certain circumstances, lead to the deformation of the carabiner. In the worst-case scenario, the forces can be so great that the carabiners might break [43].

Cross-axis load

Unfortunately, many climbers do not know that a carabiner with a load on its cross axis has less than half the breaking strength of a carabiner with a load on its longitudinal axis. D-shaped carabiners are designed to slide back into the correct position. However, this is only possible if the loop in the knot is wide enough and the carabiner can move freely. To prevent stress on the cross axis, some HMS carabiners have an internal positioning element for the lashing point. This prevents the carabiner from slipping and always ensures correct orientation and maximum breaking strength [44].

Load on one edge

The possibility of a carabiner being stressed on an edge must always be avoided. If a screw, nut or guard is incorrectly positioned, it could lead to a carabiner being applied directly to a stone ledge. This can have dramatic results. By extending it with a knot in a knot etc. one can make sure that it locks freely and away from any edge so that the stress is applied longitudinally where it is most resistant.

Load with the locking element open

The carabiner must always be closed correctly in order to have maximum breaking strength. If the rope is caught by the tip of the carabiner or something prevents the closure of the locking element, the carabiner will have well below 50% of its normal breaking strength. In the event of extreme force, the carabiner might bend or break.

Abrasion and burrs

Just like other equipment, carabiners begin to show signs of wear over time. Carabiners are personal protective equipment for mountain sports and more. Therefore, they must be carefully checked. Direct contact with other metal elements, but also the wear caused by the rope could create sharp edges or burrs. They can damage the rope or even cause it to break. For this reason, the use of carabiners with such signs of wear must be discontinued. In addition, carabiners that show signs of wear are not as strong. Abrasion may have a significant effect on decreased strength [45].

Low temperatures

During mountaineering or ice climbing, carabiners are subjected to extreme conditions. Carabiners with wire locking elements are less susceptible to frost, so they are better suited for use in such conditions.

Salt water or another corrosive agent

In areas such as the naval or oil industry, the main corrosive agent with which carabiners come into contact is salt water. Its action on the materials from which the connectors are made decreases the tensile strength and durability properties and increases the risk of accidents. Therefore, by improving corrosion resistance considerably any failure of the personal protective equipment is less likely [.

References

[1]   C. Bejinariu, D.-C. Darabont, E.-R. Baciu, I.-S. Georgescu, M.-A. Bernevig-Sava, C. Baciu, Considerations on Applying the Method for Assessing the Level of Safety at Work. Sustainability 9 (2017) 1263. https://doi.org/10.3390/su9071263

[2]   C. Bejinariu, D.C. Darabont, E.R. Baciu, I. Ionita, M.A.B. Sava, C. Baciu, Considerations on the Method for Self Assessment of Safety at Work. Environ. Eng. Manag. J. 16 (2017) 1395–1400. https://doi.org/10.30638/eemj.2017.151

[3]   British Standards Institution, BS 360, Personal protective equipment against falls from a height — Retractable type fall arresters, BSI Standards Limited, 2002.

[4]   W. Ellis, Fall Arrest Equipment, Health and Safety International Magazine/Working at height Health and Safety International Magazine, Bomel Ltd, Falls from height - Prevention and risk control effectiveness. Health and Safety Executive Research Report 116, HSE Books, 2003.

[5]   British Standards Institution, BS 365, Personal protective equipment against falls from a height — General requirements for instructions for use, maintenance, periodic examination, repair, marking and packaging, BSI Standards Limited,

2004.

[6]   British Standards Institution, BS 363: Personal fall protection equipment -
        Personal fall protection systems, BSI Standards Limited, 2008.

[7]   Directive 2001/45/EC of the European Parliament and of the Council of 27 June
        2001 amending Council Directive 89/655/EEC concerning the minimum safety
        and health requirements for the use of work equipment by workers at work. OJ L
        195, 46, 19.07.2001.

[8]   S. Wearing, L. Peebles, D. Jefferies, K. Lee, E. Ebenezer Anjorin, System
        Concepts Limited for the Health and Safety Executive, First evaluation of the
        impact of the work at height regulations, First evaluation of the removal of the
        'two metre rule'. HSE Books, 2007.

[9]   Council Directive 89/686/EEC of 21 December 1989 on the approximation of the
        laws of the Member States relating to personal protective equipment. OJ L 399,
        18-38, 30.12.1989.

[10]   D.C. Darabont, A.E. Antonov, C. Bejinariu, Key elements on implementing an
        occupational health and safety management system using ISO 45001 standard, in:
        I. Bondrea, C. Simion, M. Inta (Eds.), 8th International Conference on
        Manufacturing Science and Education (Mse 2017) - Trends in New Industrial
        Revolution. E D P Sciences, Cedex A, p. UNSP 11007, 2017.
        https://doi.org/10.1051/matecconf/201712111007

[11]   D.-C. Darabont, R.I. Moraru, A.E. Antonov, C. Bejinariu, Managing new and
        emerging risks in the context of ISO 45001 standard. Qual.-Access Success 18
        (2017) 11–14.

[12]   Centuri siguranta tip ham pentru lucru la înălțime (2018-11-10). Information on:
        http://www.mondo romania.ro/index.php?p=produse&idmeniu=8&idgrup=35
        (Accessed: November 7, 2020).

[13]   Climbing Dictionary & Glossary. MountainDays.net. Archived from the original
        on 2007-01-03. Retrieved 2006-12-05.

[14]   C. Bejinariu, D. P. Burduhos-Nergis, N. Cimpoesu, M. A. Bernevig-Sava, S. L.
        Toma, D. C. Darabont, C. Baciu, Study on the anticorrosive phosphated steel
        carabiners used at personal protective equipment, Quality-Access to Success 20(1)
        (2019) 71-76.

[15]   Online Etymology Dictionary. Information on:
        https://www.etymonline.com/search?q=carabineer (Accessed: November 03,

2018).

[16]   UIAA 121 Mountaineering and Climbing Equipment – Connectors. Union
       Internationale des Associations d'Alpinisme, 2004.

[17]   C.M. Bright, A History of Rock-Climbing Gear Technology and Standards.
       Mechanical Engineering Undergraduate Honors Theses, United States of America,
       2014.

[18]   J.D. Gorby, J. Gorby, The Stettner way: The life and climbs of Joe and Paul
       Stettner. Colorado Mountain Club, ISBN 0972441301, Colordon United States of
       America, 2003.

[19]   J. Waterman, Cloud dancers: portraits of north american mountaineers. AAC
       Press, ISBN: 978-0930410544, Colorado, United States of America, 1997.

[20]   G. Rey, 1919.Alpinisme acrobatique, Dardel, Chambery, Franța.

[21]   M. Samet, Climbing Dictionary: Mountaineering Slang, Terms, Neologism &
       Lingo: An illustrated reference. Mountaineers Books, ISBN: 978-1594855023,
       Seattle, United States of America, 2011.

[22]   The history of carabiners. Information on: https://gallantry.com/blogs/journal/history-
       of-carabiners (Accessed: September 02, 2020)

[23]   B. Gaines, Rappeling: Rope descending and ascending skills for climbing, caving,
       canyoneering and rigging (How to climb series), Falcon Guides, ISBN
       0762780800, United States of America, 2013.

[24]   Sporturi de iarna, alpinism sportiv, Serrai din sottoguda (2018 11 24). Information
       on: https://pxhere.com/ro/photo/1198580 (Accessed: November 24, 2018).

[25]   My clean- alpinism utilitar (2018 11 24). Information on: https://myclean.ro/alpinism-
       utilitar-2 (Accessed: November 24, 2018).

[26]   Carabiner Rope access Aluminium Petzl Climbing Harnesses (2020 09 02).
       Information on: https://www.pngflow.com/en/free-transparent-png-msksd
       (Accessed: September 02, 2020). https://doi.org/10.36548/jitdw.2020.3

[27]   Tree works (2020 09 02). Information on: https://www.falllineforestry.ca/tree-
       works/ (Accessed: September 02, 2020).

[28]   RRT: Rope Rescue Technician, CSFM (2018 11 24). Information on:
       https://www.code3rescue.com/rescue-courses/technical-rope-rescue (Accessed:
       November 24, 2018).

[29]   Air balloon mockup (2020 09 02). Information on: https://yellowimages.com/stock/air-

balloon-mockup-63378 (Accessed: September 02, 2020).

[30] NYC Local Law 196 Construction Safety Training Now in Effect (2018 11 24). Information on: https://www.milrose.com/insights/locallaw196 (Accessed: November 24, 2018).

[31] Injured acrobats file lawsuit connected to 2014 circus accident (2018-11-24). Information on: http://www.providencejournal.com/article/20160502/NEWS/16050994 3 (Accessed: November 24, 2018).

[32] Caving, speleology (2018-11-24). Information on: https://besthqwallpapers.com/other/cave-pit-olimp-croatia-caving-speleology-5384 (Accessed: November 24, 2018).

[33] Industrial rope access (2018 11 24). Information on: https://jasscan.com/marine/industrial-and-commercial-rope-access (Accessed: November 24, 2018).

[34] Information on: https://www.hse.gov.uk/ (Accessed: September 02, 2020). https://doi.org/10.36548/jitdw.2020.3

[35] D.P. Burduhos Nergiş, C. Nejneru, D.C. Achiţei, N. Cimpoieşu, C. Bejinariu, Structural Analysis of Carabiners Materials Used at Personal Protective Equipments, Euroinvent ICIR IOP Conference Series: Materials Science and Engineering 374(1) 012040, 2018. https://doi.org/10.1088/1757-899X/374/1/012040

[36] S. Lehner, V. Senner. Evaluation of ergonomics of a new effort saving via-ferrata carabiner-child vs. adult use. Procedia Engineering 60 (2013) 319-324. https://doi.org/10.1016/j.proeng.2013.07.026

[37] M.R.M. Aliha, A. Bahmani, S. Akhondi, Fracture and fatigue analysis for a cracked carabiner using 3D finite element simulations. Strength of Materials, 47(6) (2015) 890-902. https://doi.org/10.1007/s11223-015-9726-z

[38] D. Secunda, Standards for Climbing, ASTM Standardization News, February 1994.

[39] British Standards Institution, BS 8437, Code of practice for selection, use and maintenance of personal fall protection systems and equipment for use in the workplace, BSI Standards Limited, 2005.

[40] K.B. Blair, D. Custer, J.M. Graham, M.H. Okal, Analysis of fatigue failure in D-shaped carabiners. Sports Eng., 8 (2005) 107–113. https://doi.org/10.1007/BF02844009

[41]   V. Scott, Design of a Composite Carabiner for Rock Climbing. Final Year Project, Mechanical Engineering, Imperial College London, England, 2008.

[42]   Elderid, Carabiner Handbook (2018 12 12). Information on: https://www.edelrid.de/en/sports/knowledge/pics/2018/Karabiner_Fibel_EN_ANS ICHT.pdf?m=1536741641 (Accessed: December 12, 2018).

[43]   K. Zafren, B. Durrer, J.P. Herry, H. Brugger, Lightning injuries: prevention and on-site treatment in mountains and remote areas: official guidelines of the International Commission for Mountain Emergency Medicine and the Medical Commission of the International Mountaineering and Climbing Federation (ICAR and UIAA MEDCOM), Resuscitation, 65 (2005) 369-372. https://doi.org/10.1016/j.resuscitation.2004.12.014

[44]   A. Nagy, J. Rohacs, Measuring and modelling the longitudinal motion of paragliders. International Council of The Aeronautical Sciences, 4 (2012) 2982-2989.

[45]   D.P. Burduhos Nergis, N. Cimpoesu, P. Vizureanu, C. Baciu, C. Bejinariu, Tribological characterization of phosphate conversion coating and rubber paint coating deposited on carbon steel carabiners surfaces, Materials today: proceedings 19 (2019) 969-978. https://doi.org/10.1016/j.matpr.2019.08.009

[46]   D.P. Burduhos-Nergis, C. Nejneru, R. Cimpoesu, A.M. Cazac, C. Baciu, D.C. Darabont, C. Bejinariu, Analysis of Chemically Deposited Phosphate Layer on the Carabiners Steel Surface Used at Personal Protective Equipments, Quality-Access to Success 20(1) (2019) 77-82.

CHAPTER 2

Materials Used to Manufacture Carabiners

2.1 Materials used to manufacture carabiners

The choice of materials is an important aspect in the carabiner manufacturing process. A good understanding of the product requirements is necessary, the main factors to consider in choosing carabiner materials are:

- weight;

- resistance to static / dynamic stress to observe the appropriate standard;

- resistance to temperatures ranging between -40 °C and > 50 °C, freeze-thaw;

- sunlight – resistance to UV;

- corrosion resistance, for example, in fire extinguishing solution or sea water;

- abrasion resistance;

- durability (seconds, hours, days, years) to observe the appropriate standard;

- economic, material costs, processing cost;

- design, ergonomics, appearance.

Almost all carabiners currently used for mountaineering are made of aluminum alloys. For situations where weight is not an important factor, such as fixed anchors, steel carabiners are occasionally used due to their high wear resistance and better tensile strength [1,2].

When used for evacuation, rescue or climbing classes, steel carabiners are used. Steel melts below 1538°C, making them ideal for evacuation and rescue operations at high temperatures. Steel is hard, durable and relatively elastic, making it crack-resistant. Steel carabiners will corrode quickly if exposed to fresh or salt water. Stainless steel is not as wear resistant but has much better corrosion resistance properties [3,4].

Carabiners made of stainless steels, alloy steels, micro-alloy steels containing boron and manganese are used in certain situations where low weight is not an important factor. In some applications, their high corrosion or wear resistance makes these materials more suitable. Some examples are cave exploration, roofing or industrial applications [5].

The only advantage of aluminum carabiners is their mass. The most common aluminum carabiners weigh about 80g, while a steel one weighs about 280g. This means a difference of 200g per carabiner. The main motivation is that for every two carabiners changed, the weight of a kit increases or decreases by about half a kg. A kit containing 5 steel carabiners will weigh about 1 kg more than a kit with aluminum carabiners.

The anodizing process, i.e. electrochemical deposition of a harder layer than the base material, confers aluminum alloy carabiners improved corrosion resistance. Aluminum alloys are forged and aged alloys, based on Al-Cu, Zn, Mg, Cr systems. The 7075-T6 is usually used for the carabiner body and for the locking element. The locking element is hollow and actuated by a spring fixed on the joint. The spring is usually made of stainless steel regardless of the material used for the rest of the carabiner. The locking element is attached to the carabiner body by a stainless-steel rivet, and the closing element is locked on top of the carabiner by a second rivet. Locking systems on carabiners are metallic, usually made of 7075 alloy or polymers, usually injection molded nylon over the locking element [6].

Most simple carabiners are made of circular wire. The C shape of the carabiner is obtained by the plastic deformation of the annealed material; relatively minor changes in cross section are obtained by cold forging with the help of pressing tools. The profile of the tip is stamped and the hole(s) obtained by drilling are riveted. After forming, the carabiner is heat treated and then sanded and polished to remove sharp edges. Locking elements and metal sleeves are made by turning, milling and drilling and they are also polished [7].

2.1.1 Aluminum alloys - Duralumin

According to Collins dictionary, duralumin is defined as an aluminum alloy with a high content of copper (3.5% -4.5%), silicon, magnesium and manganese [8]. Duralumin is a durable and light alloy. It is also reflective. It is a malleable material, being easily deformed. It has good thermal and electrical conductivity. It reacts with oxygen to form aluminum oxides, which ensures its corrosion resistance. In general, duralumin alloys are soft, ductile, embossed or forged, and are available in a variety of shapes. It has high mechanical strength, and it is weldable under special conditions [9, 10].

Applications

Duralumin is used to manufacture wire, bars and rods. It is also used to manufacture wheels, plates, accessories and parts for aircraft, jet engine rotors and in the army for ammunition [11]. Carabiners have been improved several times over the decades, adjusting their mass, reliability and durability.

2.1.2   Carbon steels

Steel is an iron-based alloy containing less than 2.11% carbon. Carbon steels (also called base steels) can be defined as steels containing, in addition to carbon, other residual elements, except those added for deoxidation (silicon or aluminum) and those added to counteract the harmful effects of sulfur (manganese or cerium) [12].

The mechanical properties of carbon steels differ depending on the carbon content of steel. They have good mechanical strength properties, are plastic and tenacious materials with considerable elasticity. Their thermal and electrical conductivity, as well as the fact that they can be processed (welded, chipped, deformed) make carbon steels a suitable choice for the manufacture of machine parts [13].

Compared to other metallic or non-metallic materials, they have high specific weight, as well as low corrosion resistance [14].

The properties of carbon steels change depending on the carbon concentration. Therefore, their strength increases with the carbon content up to 1%, after which it decreases due to the amount of secondary cementite that is generated. Their necking, resilience and elongation properties decrease with increasing carbon content, while their hardness increases [15].

In addition to their mechanical properties, the technological properties of carbon steels are influenced by the percentage of carbon. Thus, the hardening capacity, castability and chip ability increase up to a certain percentage of carbon, while the properties of weldability and deformability are reduced [12].

Applications

Low carbon steels are usually obtained in rolled sheets and profiles, and they are used to build ships, vehicle bodies and household appliances. It is also known as 'wrought iron', since it is widely used for fences, gates and railings.

Medium carbon steels are easier to use in structural applications, and the quality of the material can be improved by adding small amounts of silicon and manganese. It is generally used in building and bridge structures, axles, gears, shafts, rails, pipes and couplings, vehicles, refrigerators and washing machines.

High carbon steel has a much better tensile strength compared to other carbon steels, but almost zero deformability and it is used to manufacture cutting tools, blades, mills, springs and high strength wires [12, 16].

Phosphate Coatings Suitable for Personal Protective Equipment   Materials Research Forum LLC
Materials Research Foundations **89** (2021)   https://doi.org/10.21741/9781644901113

2.1.3 Stainless steels

Stainless steels are iron-carbon alloys with a chromium content of at least 10.5% and a carbon content of up to 1.2% [17]. They are used for their corrosion resistance, which increases with increasing chromium content, and molybdenum additives increase their corrosion resistance when they come in contact with acids and chlorine solutions. Thus, there are many classes of stainless steel with variable chromium and molybdenum content that are used in different environments. Its corrosion resistance, low maintenance costs and shine make stainless steel an ideal material for many applications where both the mechanical strength of the steel and its corrosion resistance are required [18-20].

Compared to carbon steels, stainless steels do not show even corrosion when exposed to highly humid environments. Carbon steel corrodes quickly when exposed to air and humidity, the resulting oxide layer on its surface is porous and brittle. As iron oxides formed on the surface of carbon steels tend to expand and move away from it, the non-oxidized surface is exposed and attacked again when in contact with air [12, 19, 21].

Applications

Stainless steel is used in a wide variety of fields, such as: architecture and construction works, in the automotive and transport industry, in the medical field, in energy and heavy industry, in the food field, etc. [22].

2.1.4 Specific characteristics of the materials used to manufacture carabiners

The choice of materials for the manufacture of connecting elements is made, taking into account the stresses to which carabiners are exposed during operation. Therefore, there are a number of factors (corrosive agents, thermal exposure range, etc.) that must be correlated with the properties of the material [23,24]. In order to better understand the choice of the material used to manufacture carabiners, the specific properties of some types of materials 'representative' of the categories they belong to, namely: 7075-T6 aluminum alloy, C45 carbon steel and X2CrNiN23- 4 stainless steel were compared [7].

Considering that in leisure mountaineering, during climbing, the user possesses a large number of carabiners, as well as other equipment, it is recommended that they use carabiners that have low specific weight, i.e. carabiners made of a material that has low density. As shown in Fig. 2.1., carabiners made of aluminum alloys are preferred in these cases.

A warning from carabiner manufacturers to users is related to the property of these materials to conduct electricity, i.e. their electrical conductivity. The user must be aware that they might get electrocuted if the carabiners come into contact with other live components. Fig. 2.2 shows the electrical resistivity values for these three materials, which are the inverse of electrical conductivity and it is good to know that when temperature is low, the electrical resistivity of the material decreases.

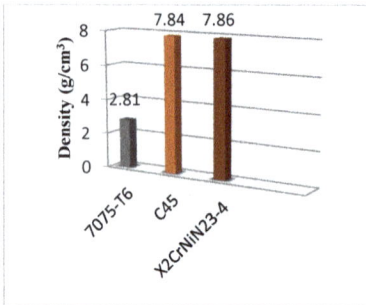

Figure 2.1 Density value.

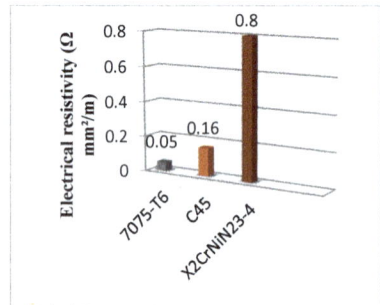

Figure 2.2 Electrical resistivity.

Another property that carabiner designers must take into account is the resistance to high temperatures and the thermal conductivity of the material. If a carabiner touches various bodies that have a high temperature (for example, when it is used in industry to take parts out of heat treatment furnaces), it is preferable that the property of metallic materials to conduct and transmit heat, thermal conductivity, be low. Figs. 2.3 and 2.4 show the thermal conductivity and thermal expansion coefficient values, which prove that, in these cases, steel is a more suitable material for the manufacture of carabiners.

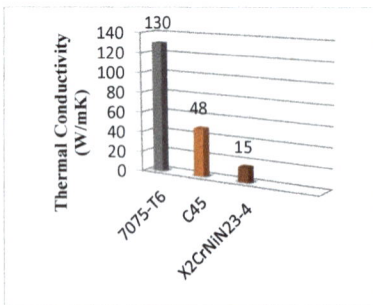

Figure 2.3 Thermal Conductivity.

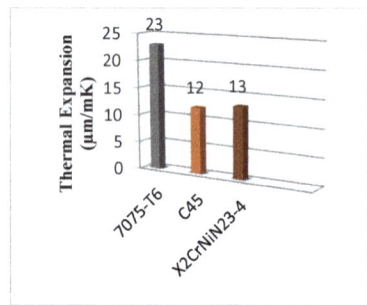

Figure 2.4 Thermal Expansion.

Materials Research Forum LLC
https://doi.org/10.21741/9781644901113

Moreover, some carabiners are used in environments where the temperature exceeds 500 °C (for example, in fire evacuation operations). Thus, Fig. 2.5 shows that, compared to aluminum alloy, steels retain their properties at high temperatures, their melting temperature exceeding 1000 °C.

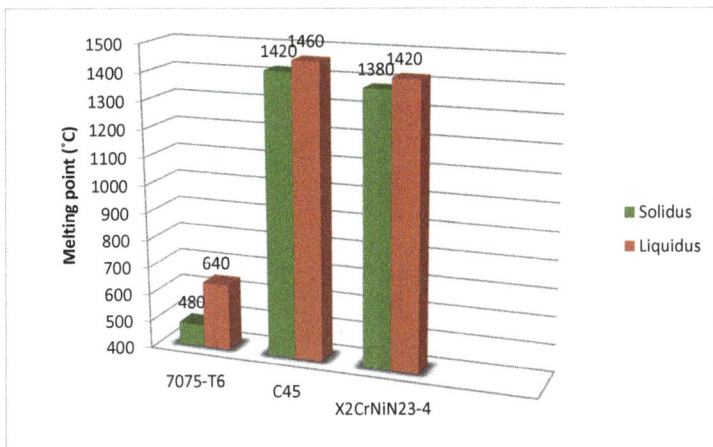

Figure 2.5 Softening and melting points carabiners materials.

The mechanical properties of materials are also important when manufacturing a carabiner. If the carabiners are hit by different bodies that are stronger and non-deformable (for example: during their use in civil and industrial engineering, carabiners may hit hard and sharp bodies, such as the edges of different equipment, scaffolding or buildings), the strength of the material from which the carabiner is made, namely its hardness, is important. Fig. 2.6 shows the hardness of the three types of materials, steels having higher hardness values than aluminum alloy.

Considering that these fasteners support heavy bodies, it is very important to know the property of the material to deform under internal or external stresses and to resume its original shape and size after the bodies have been removed, in other words, its elasticity. Fig. 2.7 shows the modulus of elasticity values for these three materials. As one may notice, aluminum alloy is the most elastic material. Therefore, in this case, following a higher load, a carabiner made of aluminum alloy may change its shape and its locking element may open accidentally, which means that the user may fall freely.

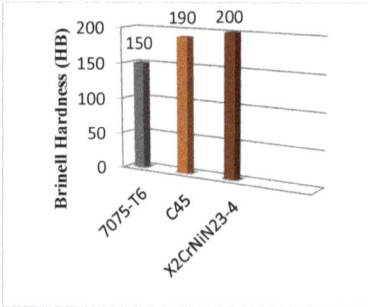

Figure 2.6 Brinell Hardness.

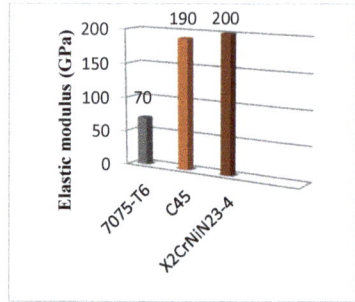

Figure 2.7 Elastic modulus.

The most important mechanical properties of the materials used to manufacture carabiners are elongation at break and tensile strength, the values of which for the three types of materials are set out in Figs 2.8 and 2.9. Under the action of external forces, some tensions arise in the mass of carabiners that oppose deformation and tear. The 12275:2013 and 362:2004 European standards impose requirements related to the tensile strength that a carabiner must have depending on its type and use.

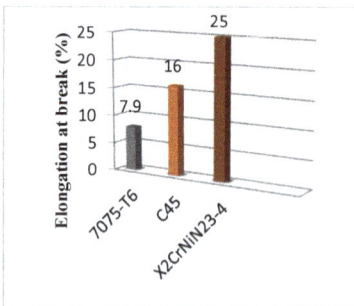

Figure 2.8. Elongation at break.

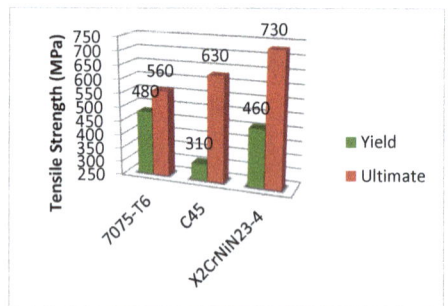

Figure 2.9. Tensile Strength

There are, however, some properties of the materials that are used to manufacture carabiners which cannot be quantified, but which must be taken into account, and they are set out in the following paragraphs.

Although aluminum alloy carabiners are lighter and easier to handle than steel carabiners, when the anchor point, cables or hooks are made of steel and the carabiner is made of aluminum, it may be affected, causing cracks to appear on its surface. In this case, it is recommended to use carabiners made of the same material as the other components [25,26].

Given the use of carabiners to also connect the textile components of fall arrest systems, the surface of the material of which the carabiner is made must be less abrasive, a quality that aluminum carabiners possess, so as not to cause premature wear of these elements [27].

An important property of the materials from which the carabiner is made is also their corrosion resistance [28,29]. Among the three materials that we compared, stainless steel has the best corrosion resistance due to the chromium content of its composition, which naturally forms on the surface of the material a passive layer of chromium oxide that protects it from corrosive agents and regenerates if the surface is scratched, while carbon steel has the lowest corrosion resistance [30].

Due to the use function and the places where the carabiner is used, after each use and before a new use, the carabiner body must be visually checked by the user. If the user notices deformations, cuts or cracks on the surface of the material that may be due to the impact of the carabiner on other bodies, the carabiner must be withdrawn from use (if no cracks are observed on the surface of the carabiner, but it has fallen from a considerable height, it must also no longer be used). In addition to blows, carabiners may be affected by contact with various corrosive substances, and if the user notices iron oxide spots on the surface of the material, carabiners cannot be used.

2.2   Carabiner manufacture process

The carabiner manufacture process includes several stages [7], namely.

1)   *Hot forging*

2)   *Surface cleaning*

Sandblasting involves introducing the product into vibrating mills or vats containing ceramic particles of various shapes and sizes. These chips remove all sharp edges, leaving the edges smooth, which is extremely important, since the finished product comes into direct contact with fixed or mobile textile elements.

3)   *Heat treatment*

This is an important stage in the manufacture of a carabiner. In order to obtain the mechanical properties necessary for the finished product, each part must be heat treated by hardening.

All steel components are heat treated by heating in closed furnaces and cooling them in the same environment, thus ensuring even hardness and tensile strength of the carabiner.

4)   *Assembly and final check*

Assembly operations are complex and require special attention. Thus, each product is checked by a specialized team in order to identify operating or manufacturing flaws.

5)   *Laser engraving*

According to the EN 12275:2013 standard, the carabiner marking must be clear, intelligible and durable, and contain at least the following information:

a) the name or trademark of the manufacturer, importer or supplier;

b) the circled letter referring to the class to which it belongs, i.e. H, K and X;

c) the minimum tensile strength value in kN, approximated to the nearest whole number smaller than the value guaranteed by the manufacturer, for the following checking modes:

- on the main axis with the locking element closed;

- on the main axis with the locking element open;

- on the secondary axis.

Markings should be in the form shown in Fig. 2.10 and bear the kN marking, either at the beginning or at the end.

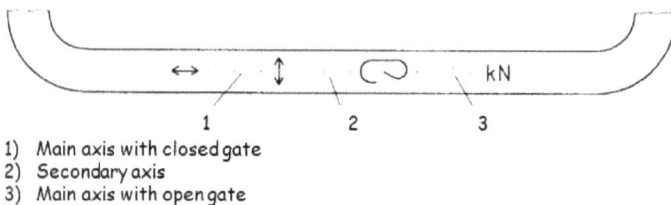

1) Main axis with closed gate
2) Secondary axis
3) Main axis with open gate

Fig. 2.10 Mark representation [20].

Before they can be marketed, carabiners go through various checking operations described below.

1) Traceability

Production is done in batches, i.e. all products manufactured in the same series remain together throughout the manufacturing process, from the moment the raw material used leaves the warehouse until the finished products pass the final checks.

2) Visual inspection

It is aimed primarily at identifying and removing all the products with visual flaws, by thorough carabiner checks.

3) Tensile test

The tensile strength of the finished products is tested in accordance with the standards in force [31,32], but some companies also introduce additional checks. This test is performed on a specific number of products in each batch.

4) CE marking and conformity of products

The products sold on the market must have an EC type check certificate. In accordance with the requirements of the European Union Personal Protective Equipment Directive 89/686/EEC [33], all products that are considered personal protective equipment must be independently checked and tested according to specific European standards.

2.3 Carabiner properties

In mountaineering, carabiners are used to connect the climber's safety rope to the anchors. In the event of a fall, they will be supported by the rope, as they reach below the level of the anchor by pulling the rope in the opposite direction. Carabiners are often sold in pairs, attached together by a nylon loop. While one carabiner is attached to the anchor, the other is attached to the rope. The characteristics that a carabiner must have can be grouped as follows: by load, by geometry and by environment [34].

The minimum load requirements for climbing connectors are set by international and European standards [31, 34]. For current carabiners made of aluminum alloy and steel, the static tests described in the EN and UIAA standards have proven that they are safe enough. Fig. 2.11 shows the tensile strength test methods according to UIAA standards and the minimum accepted strength [34].

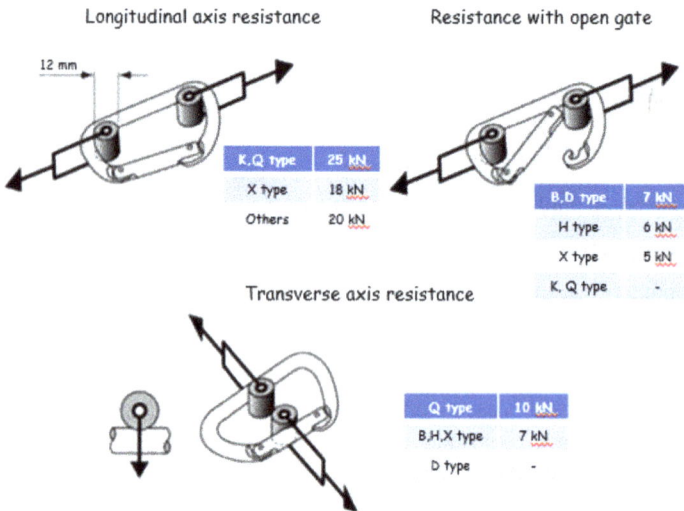

Fig. 2.11 Carabiners testing according to UIAA standards, highlighting the minimum tensile strength [34].

During normal use of carabiners, it is highly likely that they will be dropped on hard surfaces. Aluminum alloy carabiners have minor prints on their surface, as well as fine scratches due to these impacts, but continue to maintain their pre-impact properties. The height of the fall can vary from 1m to hundreds of meters, although it is unlikely that someone would reuse a carabiner after falling from distances of several tens of meters. Defects due to similar blows may occur in a steel carabiner. These damages may be internal and not necessarily visible to the naked eye, requiring non-destructive testing, such as ultrasound testing [35].

The characteristics of the environment are important, because humidity absorption and temperature variation may have significant effects on carabiner resistance. They may be exposed to very wide temperature ranges, from -40°C (Everest peak) to + 80°C due to friction during descent. In addition, it is relevant to know the effects of environment pH changes, of UV light, as well as the general chemical resistance properties of the material that the carabiners are made of [34, 36].

Due to the interactions between the rock-climbing elements and the protective equipment there are requirements regarding the geometry and design of the carabiners. Some of these, set out in EU and UIAA standards, are shown in Figs 2.12.

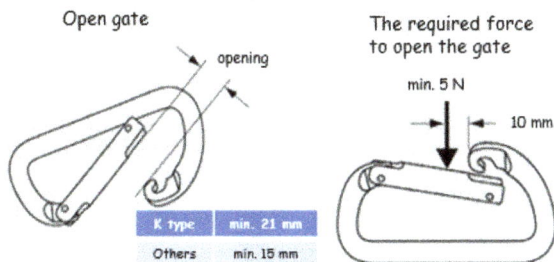

Fig. 2.12 Design requirements set by UIAA standards [34].

Some metallic materials used to manufacture carabiners are also tested for tensile stresses in a corrosive environment, thus developing surface cracks, a phenomenon known as stress corrosion. The EN standard defines the size and geometric requirements, which allow carabiners to function properly when used in safety systems, yet these specifications are not mandatory but only recommended. For instance, it is recommended that a carabiner be designed so that, when it is under a load, most of the force is borne by the side of the body opposite the locking element (Fig. 2.12).

Carabiners inevitably have lower resistance on the side of the locking element due to the stress concentrators in the fasteners of the locking element with the body. By placing the load on the opposite side of the locking element, the load to be borne by the weaker side is reduced. Moreover, this requires smaller bending of the body, so the value of the load that the carabiner withstands is close to the value of the tensile strength of the material from which the body is made [37]. There are also no standards that impose the shape of the surface on which the rope must slide, even if this is an important carabiner feature. If the radius of curvature is too small, there is a risk of the rope breaking. The UIAA standard recommends a radius of curvature of at least 4.5 mm with a contact angle of at least 120°. Carabiners must be compatible with their anchors and restraints (the climbing rope passes through the fasteners which act as a friction-based brake with the rope, so that if the climber falls, their partner can stop their fall by locking the device), but the standards do not include recommendations about these things.

Carabiners are generally compatible with stop devices and anchors, the exception being carabiners that have a large diameter in cross section and do not fit through the hole of some types of pitons (piton is a metal stake that can be fixed with a hammer in the cracks of rocks to provide an anchorage point for climbers). Another feature that carabiners must possess is their easy hand use and handling, so that the climber be able to catch the rope while in a difficult situation [34]. Table 2.1 contains a summary of the above design specifications.

Table 2.1 The main factors influencing the design of the carabiner.

Load	Environment	Geometry
The main axis	Temperature	Rope contact surface
Secondary axis	Coefficient of friction	Rope diameter
Opening the gate	Humidity	Anchor constraints
The impact	Chemical resistance	Compatibility with other components
Residual resistance	Corrosion and wear resistance	User handling
	UV resistance	

Following the critical review of literature on the failure of personal protective equipment against falls from a height, due to damage to the fasteners, it was found that carabiner durability was affected by several factors that take into account the various stresses, environment and their geometry. Thus, it was noticed that the decommissioning of carabiners made of carbon steel was due to the appearance of iron oxides spots on their surface or as a result of suspicions regarding the occurrence of internal cracks. Therefore, the study addresses the improvement of the corrosion resistance of carbon steel used to manufacture carabiners, as well as the development of a method to protect them against mechanical shocks by depositing elastomer-based coatings.

Considering that carabiners made of carbon steel are used in areas such as: navigation, rescue/evacuation operations, oil industry, construction works, etc., improving the properties of corrosion resistance and protecting carbon steel against mechanical shocks bring about new opportunities for the development of personal protective equipment, thus providing better safety to users.

References

[1]   D. Harutyunyan, G.W. Milton, T.J. Dick, J. Boyer, On ideal dynamic climbing ropes, Proceedings of the Institution of Mechanical Engineers. Part P: Journal of Sports Engineering and Technology 231(2) (2017) 136-143. https://doi.org/10.1177/1754337116653539

[2]   D.P. Burduhos Nergiş, C. Nejneru, D.C. Achiţei, N. Cimpoieşu, C. Bejinariu, Structural Analysis of Carabiners Materials Used at Personal Protective Equipments, Euroinvent ICIR IOP Conference Series: Materials Science and Engineering 374(1) (2018) 012040. https://doi.org/10.1088/1757-899X/374/1/012040

[3]   R. Kornijów, A. Drgas, K. Pawlikowski, The experimental set for in situ research of benthic communities in marine and freshwater ecosystems. Knowl. Manag. Aquat. Ecosyst. 418 (2017) 12. https://doi.org/10.1051/kmae/2017003

[4]   D.P. Burduhos-Nergis, C. Bejinariu, S.L. Toma, C.A. Tugui, E.R. Baciu, Carbon steel carabiners improvements for use in potentially explosive atmospheres, 2020 SESAM MATEC Web of Conferences 305 (2020) 00015. https://doi.org/10.1051/matecconf/202030500015

[5]   B. Kane, H.D. Ryan, Residual strength of carabiners used by tree climbers. Arboriculture & Urban Forestry 35(2) (2009) 75-79.

[6]   M. May, S. Furlan, H. Mohrmann, G.C. Ganzenmüller. To replace or not to replace?-An investigation into the residual strength of damaged rock climbing safety equipment. Engineering Failure Analysis 60 (2016) 9–19. https://doi.org/10.1016/j.engfailanal.2015.11.036

[7]   D.P. Burduhos-Nergis, C. Baciu, P. Vizureanu, N.M. Lohan, C. Bejinariu, Materials types and selection for carabiners manufacturing: A review. In Proceedings of the IOP Conference Series: Materials Science and Engineering; Institute of Physics Publishing; 572, 2019. https://doi.org/10.1088/1757-899X/572/1/012027

[8]   Collins English Dictionary (2019-01-26). Information on: https://www.collinsdictionary.com/dictionary/english/duralumin (Accessed: January 26, 2018).

[9]   L. Persson Erik, Aluminum Alloys: Preparation, Properties and Applications. Materials Science and Technologies. Nova Science Pub Inc., ISBN: 978-1611223118, New York, United States of America, 2011.

[10]   P.-E. Nica, M. Agop, S. Gurlui, C. Bejinariu, C. Focsa, Characterization of

Aluminum Laser Produced Plasma by Target Current Measurements. Jpn. J. Appl. Phys. 51 (2012) 106102. https://doi.org/10.1143/JJAP.51.106102

[11] Engineering Division McCook Field, Structural Analysis and Design of Airplanes. Wexford College Press, ISBN 78-1929148462, Minot, United States of America, 2005.

[12] I. Alexandru, A. Alexandru, V. Cojocaru, I. Carcea, G. Palosanu C-tin. Baciu, V. Bulancea, M. Calin, R. Popovici, Alegerea şi utilizarea materialelor metalice. Ed. Didactică şi Pedagogică, ISBN 973-30-5549-2, Bucureşti, România, 1997.

[13] Carbon Steel Handbook. EPRI, Palo Alto, CA: 1014670, 2007.

[14] N. Labjar, L. Mounim, B. Fouad, N.E. Chihib, S.E. Hajjaji, J. Charaf, Corrosion inhibition of carbon steel and antibacterial properties of aminotris-(methylenephosphonic) acid. Materials Chemistry and Physics, 119 (2010) 330-336. https://doi.org/10.1016/j.matchemphys.2009.09.006

[15] J.E. Bringas, Handbook of comparative world steel standards. W. Conshohocken, PA: ASTM International, 2004.

[16] I. Chesa, N. Lascu Simion, C. Mureseanu, C. Rizescu, M.S. Teodorescu, Mărci şi produse din oţel, Ed. Tehnica, ISBN 973-31-0027-7, Bucureşti, România, 1989.

[17] The International Organization for Standardization, ISO 15510, Stainless steels - Chemical composition, Switzerland, 2014.

[18] P. Marshall, Austenitic Stainless Steels. Microstructure and mechanical properties. Springer Netherlands, ISBN: 978-0-85334-277-9, Heidelberg, Germany, 1984.

[19] D. Peckner, I.M. Bernstein, Handbook of Stainless Steels. McGraw Hill, ISBN: 978-0070491472, New York, United States of America, 1977.

[20] M. Hadji, R. Badji, Microstructure and mechanical properties of austenitic stainless steels after cold rolling. Journal of Materials Engineering and Performance, 11(2) (2002) 145 – 151. https://doi.org/10.1361/105994902770344204

[21] American Society for Testing and Materials, Advances in the technology of stainless steels and related alloys, ASTM Special Technical Publication, nr. 369, 1965.

[22] N.R. Baddoo, Stainless steel in construction: A review of research, applications, challenges and opportunities. Journal of Constructional Steel Research, 64(11) (2008) 1199-1206. https://doi.org/10.1016/j.jcsr.2008.07.011

[23] C. Bejinariu, D.-C. Darabont, E.-R. Baciu, I.-S. Georgescu, M.-A. Bernevig-Sava, C. Baciu, Considerations on Applying the Method for Assessing the Level of Safety at

Work. Sustainability 9 (2017) 1263. https://doi.org/10.3390/su9071263

[24] D.C. Darabont, A.E. Antonov, C. Bejinariu, Key elements on implementing an occupational health and safety management system using ISO 45001 standard, in: Bondrea, I., Simion, C., Inta, M. (Eds.), 8th International Conference on Manufacturing Science and Education (Mse 2017) - Trends in New Industrial Revolution. EDP Sciences, Cedex A, p. UNSP 11007, 2017. https://doi.org/10.1051/matecconf/201712111007

[25] C. Bejinariu, D.P. Burduhos-Nergis, N. Cimpoesu, M.A. Bernevig-Sava, S.L. Toma, D.C. Darabont, C. Baciu, Study on the anticorrosive phosphated steel carabiners used at personal protective equipment, Quality-Access to Success 20(1) (2019) 71-76.

[26] D.P. Burduhos-Nergis, A.V. Sandu, D.D. Burduhos-Nergis, D.C. Darabont, R.-I. Comaneci, C. Bejinariu, Shock Resistance Improvement of Carbon Steel Carabiners Used at PPE, MATEC Web Conf. 290 (2019) 12004 https://doi.org/10.1051/matecconf/201929012004

[27] D.P. Burduhos Nergis, N. Cimpoesu, P. Vizureanu, C. Baciu, C. Bejinariu, Tribological characterization of phosphate conversion coating and rubber paint coating deposited on carbon steel carabiners surfaces, Materials today: proceedings 19 (2019) 969-978. https://doi.org/10.1016/j.matpr.2019.08.009

[28] D.P. Burduhos-Nergis, C. Nejneru, D.D. Burduhos-Nergis, C. Savin, A.V. Sandu, S.L. Toma, C. Bejinariu, The Galvanic Corrosion Behavior of Phosphated Carbon Steel Used at Carabiners Manufacturing, Revista de chimie 70(1) (2019) 215-219. https://doi.org/10.37358/RC.19.1.6885

[29] D.P. Burduhos-Nergis, P. Vizureanu, A.V. Sandu, C. Bejinariu, Evaluation of the Corrosion Resistance of Phosphate Coatings Deposited on the Surface of the Carbon Steel Used for Carabiners Manufacturing, Applied Sciences 10(8) (2020) 2753. https://doi.org/10.3390/app10082753

[30] D.P. Burduhos-Nergis, C. Nejneru, R. Cimpoesu, A.M. Cazac, C. Baciu, D.C. Darabont, C. Bejinariu, Analysis of Chemically Deposited Phosphate Layer on the Carabiners Steel Surface Used at Personal Protective Equipments, Quality-Access to Success 20(1) (2019) 77-82.

[31] British Standards Institution, EN 12275, Mountaineering equipment – Connectors – Safety requirements and test, BSI Standards Limited, 2013.

[32] British Standards Institution, BS 362, Personal protective equipment against falls from a height — Connectors, BSI Standards Limited, 2004.

[33] Council Directive 89/686/EEC of 21 December 1989 on the approximation of the laws of the Member States relating to personal protective equipment. OJ L 399, 30.12.1989, 18-38.

[34] V. Scott, Design of a Composite Carabiner for Rock Climbing. Final Year Project, Mechanical Engineering, Imperial College London, England, 2008.

[35] UIAA 121 Mountaineering and Climbing Equipment – Connectors. Union Internationale des Associations d'Alpinisme, 2004.

[36] British Standards Institution, BS 364, Personal protective equipment against falls from a height — Test methods, BSI Standards Limited, 1993.

[37] A.B. Spieringsa, O. Henkelb, M. Schmida, Water absorption and the effects of moisture on the dynamic properties of synthetic mountaineering ropes. International Journal of Impact Engineering, 34 (2007) 205–215. https://doi.org/10.1016/j.ijimpeng.2005.08.008

[38] British Standards Institution, BS 362, Personal protective equipment against falls from a height — Connectors, BSI Standards Limited, 2004.

CHAPTER 3

Objectives and Methodology of Experimental Research

Over time, carabiners have undergone a series of changes that have led to improved user safety when working at a height. Although connectors of this type made of aluminum are the most common, especially among leisure climbers, due to its low density, carbon steel remains one of the most used materials for the manufacture of carabiners when greater tensile strength or high temperature resistance is required.

Considering the diversity of areas in which carabiners are used, the improved corrosion and impact resistance properties of carbon steel, in addition to its other properties and its low price, lead to these carbon steel connectors providing increased safety and thus be suitable in as many environments as possible.

The experimental research conducted in the study focused on obtaining and characterizing a phosphate coat designed to improve the corrosion resistance properties of carbon steel, as well as its subsequent use as a substrate for elastomer-based paint coating. This paint has the role of absorbing the shock when the carabiner hits another body or falls from a height.

3.1 General objectives

Due to the high number of occupational accidents caused by falling from a height, the attempt to reduce them by any improvement of personal protective equipment, intended for people working at a height, is a necessity.

Although carbon steel carabiners have high mechanical properties, they have some weaknesses. According to the EN 362:2004 standard, carabiners on the surface of which rust spots can be seen are no longer safe to use. Also, if carabiners hit another object, micro-cracks may occur inside the material that decrease its properties. In these cases, the connector should not be used any more.

The main goal is to improve the corrosion and impact resistance properties of the carbon steel used to manufacture carabiners. A number of steps must be taken to achieve this goal:

(i)   analysis of the raw material used to manufacture carabiners, to determine its chemical composition and structural characterization;

(ii) sample preparation depending on the tests that will be performed on them, and phosphating process solution preparation;

(iii) sample phosphating and subsequent deposition of a coat of elastomer-based paint;

(iv) coat characterization by various laboratory tests in order to identify the properties obtained.

Considering the main goal of this study, three types of phosphating solutions with different substances and concentrations were prepared to improve the corrosion resistance properties of carbon steel. The resulting coats were characterized and compared. The chemical composition of the phosphating solutions had the STAS 7969-85 standard, entitled 'Phosphating solutions', as starting point, but it was modified by the experiment. Following the comparison of the resulting phosphate coats, a safe phosphating solution was chosen on which the elastomer-based paint will be later deposited.

The schedule of this research, detailed in Table 2.1, was developed and implemented in order to meet the main goal, which aims to improve the corrosion and impact resistance properties of carbon steel used to manufacture carabiners.

Table 3.1 Experimental research program.

The state of art
The first stage of this research was the study of bibliographic materials about carabiners, as well as the materials from which they are made. The analysis of the problems that appear when using carabiners made of various materials led to the study of a possibility to improve the properties of carbon steel used in their manufacture.

Analysis of the raw material
The sample that was investigated was taken from a carabiner purchased from the CAMP company. It was analyzed from a structural and chemical composition point of view.

Sample preparation
Following the purchase of a material equivalent to the steel used in the manufacture of carabiners, it was cut in order to obtain samples with the specific dimensions required for each test. The next step in preparing the samples was sanding and polishing them.

Material phosphating
This stage consists in the preparation of the solutions used in the phosphating process, as well as the actual phosphating of the samples prepared at the previous stage.

Samples painting
Following phosphating, a number of samples were selected and a layer of elastomer-based paint was deposited on their surface.

Layers characterization	The study of the properties of the deposited layers was performed by laboratory investigations aimed to determining the improved properties of carbon steel by: chemical composition analysis, structural and mechanical characterization, as well as corrosion resistance.
Results interpretation	The experimental values were represented graphically, and then were analyzed and interpreted according to the existing data in the literature. In some cases, for example, corrosion resistance, results were compared with the values of the base material.
Conclusions	The obtained results were synthesized, being expressed concisely in the last stage of the research. Also at this stage, personal contributions were highlighted, as well as future research perspectives.

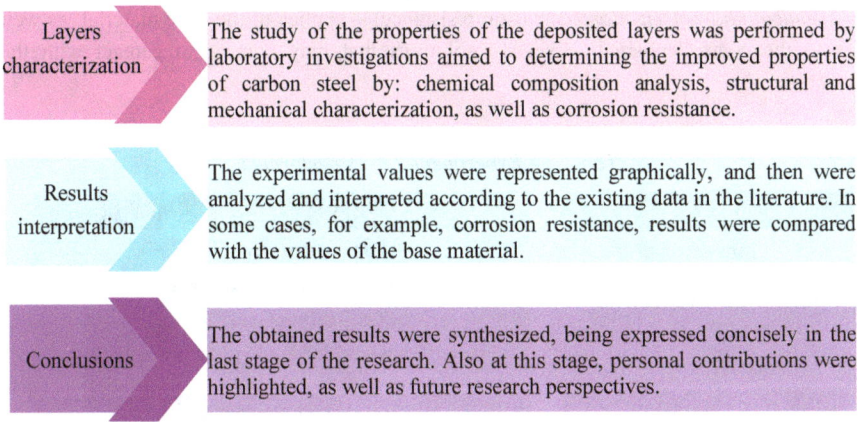

The experimental research schedule helps to achieve the main objective of the research, generating a series of results that have a significant contribution to the field of materials.

3.2 Experimental research methodology

Addressing this topic in this book contributes to the improvement of the properties of carbon steel carabiners. This reduces the disadvantages of carbon steel, as well as reducing the user's risk of falling from a height due to its low corrosion resistance and impact resistance properties.

The experimental tests conducted in this research are shown in Table 2.2 and aim to characterize the resulting phosphate coats, as well as the paint coat subsequently deposited by chemical, structural, mechanical analysis.

The following laboratory research was performed:

- *elemental analysis* - it is necessary to determine the chemical composition of the base material of which the carabiner is made, as well as the composition of the coats subsequently deposited;

- *structural characterization* – is carried out in order to study the microstructure of the metal and the formation of crystals by phosphating;

- *mechanical characterization* – helps to determine the coefficient of friction and modulus of elasticity of phosphate and paint coats, as well as to quantify the energy absorbed following the impact;

- *chemical characterization* (corrosion resistance) – applies to the base material, as well as to the coats deposited in several solutions that may come into contact with the carabiner;

Table 3.2 Experimental tests performed.

Elemental analysis	X-ray spectroscopy by energy dispersion (EDX)
Structural characterization	Optical microscopy Scanning electron microscopy (SEM) X-ray diffraction (XRD) Fourier Transform Infrared Spectroscopy (FTIR)
Mechanical characterization	Scrach test and microindentation test Încercarea la impact
Corrosion resistance	Liniar and cyclic polarization Electrochemical impedance spectroscopy

3.2.1   Material used in experimental research

Due to the operating conditions, the materials used to manufacture carabiners must have good properties of tensile strength, corrosion resistance and impact resistance (resilience) [1].

The material of a type X carabiner (Fig. 3.1), used for attachment to the piton, was structurally analyzed to determine what carabiners are made of. According to the data provided by the manufacturer, it has a minimum breaking strength of 22 kN and a mass of 175g, and it is made of carbon steel. The metallographic research on the carabiner material was carried out by cutting a section and turning it into a metallographic sample.

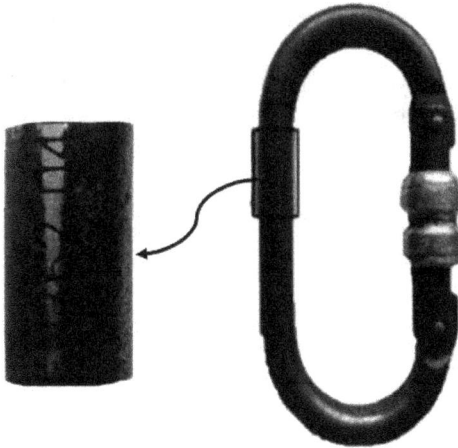

Figure 3.1 X type carabiner.

A number of steps were taken to prepare the metallographic sample for analysis:

(i)   the sample was cut off using a Microcut 150 grinder;

(ii)   sample embedding was done in order to make its processing easier. The sample was hot embedded in the resin;

(iii) the mechanical grinding of the sample was performed using metallographic papers. This consisted of a series of operations starting with high-grain grinding paper of 150 particles/mm^2 and finishing off with small-grain metallographic paper of 1000 particles/mm^2. At the end of the grinding process, the sample was rinsed under running water to remove impurities and then dried by wiping;

(iv)   the sample was polished mechanically, obtaining a mirror gloss surface. In order to achieve this surface, the sample was polished on felt, using aluminum oxide as polishing agent. The sample was again rinsed under running water, degreased with alcohol and dried by dabbing on filter paper;

(v)   the metallographic attack was performed in order to highlight the structural constituents. Since this carabiner is made of steel, the attack was carried out with a solution of 4% nitric acid dissolved in ethyl alcohol, called nital.

The chemical composition of the material which the purchased carabiner was made of was determined using a Foundry Master optical emission spectrometer equipped with

WASLAB software from the Department of Technologies and Equipment for Materials Processing, Faculty of Materials Science and Engineering, 'Gheorghe Asachi' Technical University of Iași, as shown in Table 3.3. According to this result, the material used to manufacture the carabiner is carbon steel.

Tabelul 3.3 Chemical composition of X carabiner material.

Element	Fe	C	Si	Mn	P	S	Ni	Cr
Percent, (%)	Balance	0.47	0.13	0.75	0.03	0.038	0.20	0.20

The microstructure of the material was determined with a MEIJI TECHNO IM7200 optical microscope and a Vega Tescan LMH II scanning electron microscope. As shown by Figs. 3.2 and 3.3, the microstructure of the material consists of ferrito-perlitic equiaxial grains with clear-cut edges and relatively even distribution of perlitic-type grains. The chemical evenness of the material structure is shown in Fig. 3.4, being determined using the EDAX detector attached to a scanning electron microscope.

In order to improve it, the material has undergone a heat hardening treatment in order to obtain high hardness, which causes an increase in wear resistance. The hardness value of the sample was determined by the Vickers method and it was 3403 MPa. Tempering was followed by a high draw-tempered stage designed to increase the toughness of the material [2].

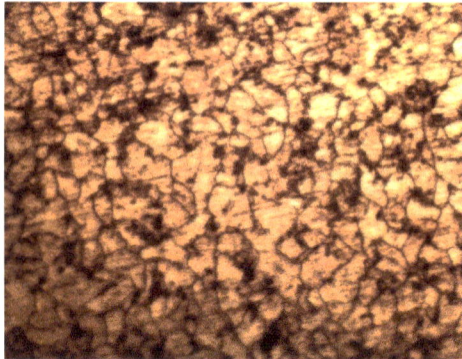

Figure 3.2 Optical microstructure, X50. Equiaxed pearlitic-ferrite grains.

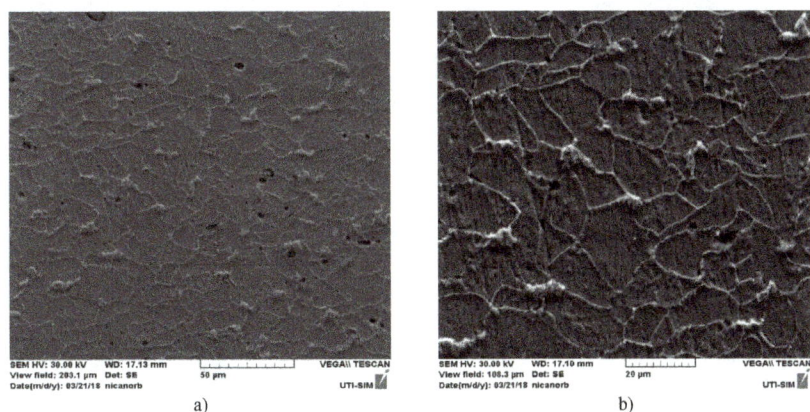

a)   b)

*Figure 3.3 The carabiner material morphology investigated at a magnification of: a) 1kx;
b) 2kx.*

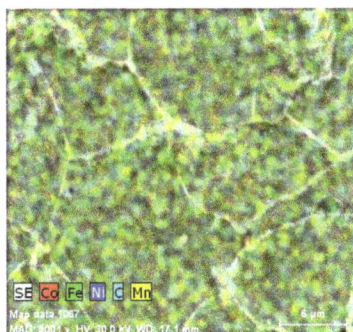

Figure 3.4 EDX microstructure of carabiner material.

Following a discussion with a representative of a well-known carabiner manufacturer, it turned out that they frequently use 1045 steel, the chemical composition of which is compliant with the SAE J1397-1992 standard shown in Table 3.4.

Therefore, following this discussion and the prior analysis of the carabiner X material, we decided to use for this research a steel equivalent to 1045 steel, i.e. C45 steel. The chemical composition of this steel compliant with the SR EN 10083-2:2006 standard is shown in Table 3.5, and the chemical composition of the steel batch used for the thesis is shown in Table 3.6.

Table 3.4 *Chemical composition of 1045 steel.*

Element	Min.	Max.
C	0.43	0.5
Mn	0.6	0.9
P	-	0.04
S	-	0.05
Si	0.1	0.25
Cu	0.2	-
B	0.0005	0.003
Pb	0.1	0.35
Fe	balance	

Table 3.5 *Chemical composition of C45 steel.*

Element	Min.	Max.
C	0.42	0.50
Mn	0.50	0.80
P		0.04
S		0.045
Si	0.17	0.37
Ni		0.30
Cr		0.30
Cu		0.30
As		0.05
Fe	balance	

Table 3.6 *The chemical composition of C45 steel used in experimental research.*

Element	Fe	C	Si	Mn	P	Cu	Cr
Percent, (%)	balance	0.45	0.22	0.98	0.02	0.15	0.17

C45 carbon steel is an all-purpose steel, which possesses good properties of hardness and tensile strength at high temperatures, its biggest disadvantage being the low corrosion resistance, a property that could be improved by coating the carbon steel carabiner surface with insoluble phosphates.

3.2.2 Phosphate coating methodology and technology

The use of phosphate coats or protection of steel surfaces has been known since the early twentieth century. During this time, much of the world's production of cars, refrigerators and furniture was treated in this way. The coating of steel with a layer of phosphate to prevent the occurrence of iron oxides was first registered by Ross, by a British patent, in 1869. In this patent, the hot steel parts were immersed in phosphoric acid. Since then, the phosphating process as well as the phosphating solutions have been continuously developed [3-5].

Carbon steel is the most commonly used material in the manufacture of parts, accounting for about 85% of annual steel production worldwide. Although carbon steel has low corrosion resistance, it is used in marine applications, chemical processing, oil production and refining, metal construction and metalworking equipment, as well as for the manufacture of carabiners used in various fields [6].

There are different ways to prevent the materials from corroding: inhibitors, coatings or anodic/cathodic protection [7,8]. Surface treatment is an effective technique for improving corrosion resistance, which includes coating by chemical conversion, electrical plating, physical vapor deposition, etc. [9-11]. Among these surface treatments, chemical conversion treatment is a simple and cost-effective method, used in a wide range of applications.

Phosphating is one of the most important conversion deposition processes, being widely used in many industries for corrosion protection, wear resistance and as a paint substrate [12].

The phosphating process may be defined as the formation of a coat of insoluble phosphates on the surface of the metal by a chemical reaction that takes place between its surface and the phosphating solution. In this coating, the phosphate layer not only covers the entire surface of the material, but also forms bonds with the base material [13-16].

One of the main stages of carabiner manufacturing is the cleaning of their surfaces. This may also be considered as a stage of preparing the surface of the carabiner for phosphating, as the structure and type of surface of the material may have a negative or positive impact on the phosphate coat. The surface of the carabiner body can be cleaned and activated by sanding and polishing or by blasting with ceramic particles [17].

In order to coat the surface of the material evenly, the phosphating process comprises several steps which are shown in Fig. 3.5. Depending on the properties of the surface on which the phosphate coat is to be deposited, as well as depending on the substances that are used to make the solutions, some steps may be added or removed [18].

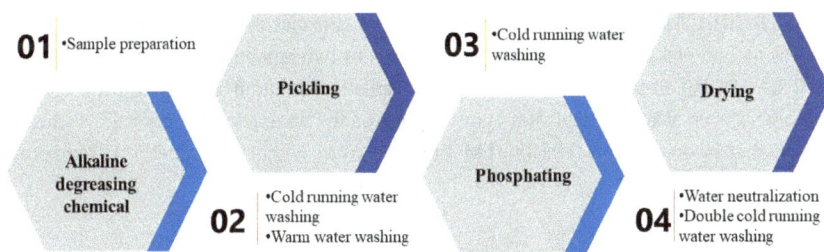

Figure 3.5 Phosphating process stages.

In order for the phosphate coat to improve the corrosion resistance of the carabiner, the surface on which the phosphate coat is to be deposited must be prepared. Therefore, before immersing the sample in the phosphating solution, it must go through two other prior surface preparation stages. After each stage, carabiners must be rinsed under running water in order to remove the chemical compounds from their surface, which have occurred after degreasing and pickling. The last stage of the phosphating process, drying, may be performed at room temperature or at high temperatures $100 \div 150 \ ^{\circ}C$, in ovens [19].

The device shown in Figure 3.6 was used for the actual conduct of the phosphating process.

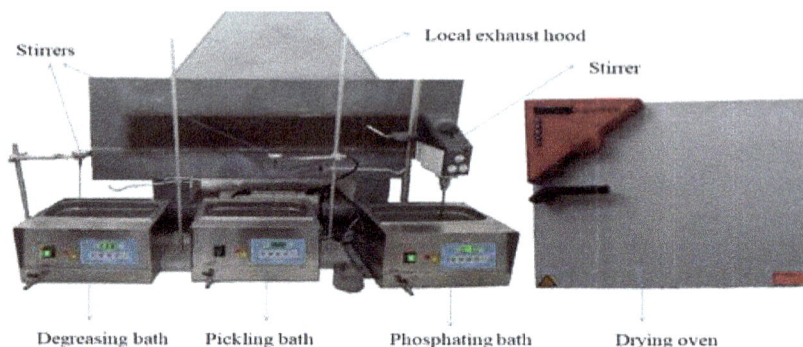

Figure 3.6 General view of phosphating installation.

The heating of the degreasing and phosphating solutions is done in digital thermostatic baths of the DIGIBATH-2 Raypa type, thus maintaining constant the working temperature. The degreasing and pickling baths are stirred by means of two agitators driven by SIEMES 1AF 2210 0A, 220V electric motors, and the phosphating solution is stirred with a R2120, Heidolph stirrer at a speed of 500 rpm. Drying of the samples after crystalline chemical phosphating is done in an APT.lineTM ED (E2) oven within the 100-150°C temperature range.

3.2.2.1 Degreasing

It is well known that carbon steel has low corrosion resistance, so it oxidizes when stored in humid places. In order to prevent the atmospheric oxidation of steel, it is coated with a thin layer of oil or grease [19].

Before being phosphated, the surface of the carbon steel sample must not contain fat particles [20]. This step is important to avoid coating the surface of the material only partially. A greasy surface may lead to uneven pickling, so only part of the surface is activated.

Degreasing is carried out in particular to remove oils or greases used during the preparation and storage of the material and to remove other residues. The duration of the degreasing stage depends on the level of contamination of the samples and they may generally be visually checked by picking up the sample from the degreasing solution at regular intervals. If an even water film occurs on the surface of the sample, degreasing may be considered completed.

The chemical composition of the alkaline chemical degreasing solution used in our research is shown in Table 3.7.

Table 3.7 The chemical composition of the degreasing solution for 2 liters.

Substance	Quantity [g]
Sodium hydroxide (NaOH)	90
Sodium carbonate (Na_2CO_3)	60
Trisodium phosphate ($Na_3PO_4 \cdot 10H_2O$)	60
Sodium silicate ($Na_2SiO_3 \cdot 9H_2O$)	10
Surfactant	13

Sodium hydroxide

Originally used to manufacture soap, sodium hydroxide is the main substance used to manufacture degreasing solutions; it is also the most widely used base in the chemical industry, being used for the manufacture of sodium salts and detergents, for pH regulation and for organic synthesis [21]. Sodium hydroxide, also known as caustic soda, is a white solid ionic compound consisting of Na^+ sodium cations and OH^- hydroxyl anions, having the chemical formula NaOH [22].

It is an alkaline base able to break down proteins at ambient temperature, causing severe chemical burns. It absorbs moisture and carbon dioxide from the air, being very soluble in water. In our case, the amount of 90 g was dissolved in 200 ml of distilled water.

Dissolving solid sodium hydroxide in water is an extremely exothermic reaction that releases a large amount of heat. The resulting solution is usually colorless and odorless.

Surfactants are also added to stabilize the dissolved substances and to prevent the deposition of fat particles in the degreasing solution.

Sodium carbonate

Sodium carbonate is an active component in many degreasing solutions and powder detergents. It is used to break down grease and oil, and to remove lubricants from the surface of metals [23].

Sodium carbonate was originally extracted from the ashes of plants growing in sodium-rich soils, and it was used to make potash. Due to this use, it is also known as baking soda. It is currently obtained from table salt through the Solvay procedure [24]. Sodium carbonate is a white, crystalline and hygroscopic powder that has the chemical formula Na_2CO_3. 60 g of sodium carbonate were dissolved in 200 ml of distilled water to obtain the final degreasing solution. Following the addition of sodium carbonate to the water, an exothermic reaction took place, forming an alkaline solution consisting of carbonate anions and hydroxyl groups.

Trisodium phosphate

One of the most important uses of trisodium phosphate is as a cleaning agent for surfaces to be coated with paint, improving its adhesion to the material. Due to its high pH (pH=12) it is used, in combination with surfactant detergent, as a component in solutions used to remove oil and grease from metal parts [25].

Trisodium phosphate is a white, inorganic compound in the form of granules or crystals having the chemical formula Na3PO4. In the form sold on the market, trisodium phosphate is partially hydrated and may be purchased from anhydrous trisodium phosphate to trisodium phosphate dodecahydrate [26]. 60 g of $Na_3PO_4 \cdot 10H_2O$ were added to regulate the acidity of the degreasing solution. It was initially dissolved in 200 ml of distilled water and a hydrolysis reaction took place, which resulted in $NaHPO_4$ and NaOH disodium phosphate.

Sodium silicate

Sodium silicate is the general name for compounds that have the chemical formula $Na_2 \cdot SiO2+x$. Sodium metasilicate was used for the degreasing solution, which is a solid, crystalline, white, hygroscopic chemical compound with the chemical formula Na_2SiO_3 [16]. Sodium metasilicate is an important component in detergents used in industry. The silicon content protects metal, glass or ceramic surfaces from corrosion and disperses oil/grease in solution, preventing it from re-depositing [27].

Due to its solubility in water, 10 g of sodium metasilicate were dissolved by an exothermic reaction in 66 ml of distilled water resulting in an alkaline solution, subsequently added to the degreasing solution to inhibit carbon steel corrosion by stopping its anodic dissolution.

Surfactant

The surfactant (detergent) is one of the main components of the degreasing solution, because when it is dissolved in a solution or in water its particles move to the interface between liquid and solid (dirt), changing its properties, thus removing the dirt particles on the metal surface [28]. 13 g of commercially available surfactant detergent were used for the degreasing solution, which was initially dissolved in 66ml of distilled water. All components of the solution were mixed and distilled water was added up to 2000 ml.

3.2.2.2 Pickling

Pickling is a method of treating the surface of the sample used to remove inorganic substances, and rust from ferrous materials or aluminum alloys. During the manufacturing or processing of the steel, a coat of oxides is formed on the surface of the sample. These impurities can affect the use or further processing of the part, such as surface plating or painting [29].

Despite the high number of methods of removing the oxide coat (abrasive sandblasting, salt baths, brushing) pickling is the most common [30]. It extends the working life of the part, evens out the surface and prepares it for depositing new coats on the sample.

The pickling rate depends on several variables: basic components of the pickled steel, type of adhesion of the oxides, concentration of acid and concentration of ferrous chlorides in the solution, temperature at which the pickling is conducted, stirring, immersion time and presence of inhibitors [31].

In the phosphating process, the pickling is performed after degreasing. The part is immersed into a solution made of the substances shown in Table 3.8 at ambient temperature for 20 minutes.

Table 3.8 The chemical composition of the degreasing solution for 2 liters.

Substance	Quantity
Hydrochloric acid (HCl)	300 ml
Hexamethylenetetramine ($C_6H_{12}N_4$)	0.9 g
Sodium sulphate (Na_2SO_4)	0.3 g

Hexamethylenetetramine

Also known as hexamine or urotropin, hexamethylenetetramine is a white, water-soluble, heterocyclic crystalline organic compound with the chemical formula $C_6H_{12}N_4$ [32].

Due to the general aggressiveness of acid solutions used in industry for acid pickling or cleaning of metal, hexamethylenetetramine is used to inhibit corrosion, due to its absorption on the metal surface by electron donation. This inhibitor blocks the active sites and increases the absorption process, reducing the corrosion rate and extending the life of the sample with which the pickling solution comes into contact [33, 34].

Before hexamethylenetetramine was added to the hydrochloric acid, 0.9 g was dissolved in 33 ml of distilled water.

Sodium sulfate

Sodium sulphate is a solid, white, water-soluble inorganic compound with the chemical formula Na_2SO_4 [35]. Sodium sulphate was used as a substitute for sulfuric acid in the pickling solution to reduce the use of acids causing great environmental damage [36]. 0.3 g of sodium sulphate was initially dissolved in 33 ml of distilled water, for the pickling solution.

Hydrochloric acid

Hydrochloric acid, also called muriatic acid, is a colorless inorganic chemical compound that has a strong odor and the chemical formula HCl [37]. It is an important chemical that is widely used. Its most common uses are: steel pickling, oil acidification, calcium chloride production and ore processing [38].

Hydrochloric acid is used in the pickling operations of carbon steel, alloy or stainless steel, to remove iron oxides, by transforming them into soluble compounds, before further processing of the steel, for example: rolling, phosphating, galvanizing, etc. [39].

Since 1964, sulfuric acid has replaced by hydrochloric acid in pickling solutions. In addition to lower heating costs, as it can be used at room temperature, pickling using hydrochloric acid is faster and more efficient because less iron salts are deposited on the surface of the pickled material and less hydrogen penetrates by diffusion. The only significant disadvantage of hydrochloric acid is its volatility, which is higher than that of sulfuric acid [40].

In order to protect the pickled metal from hydrochloric acid, an inhibitor is added to the pickling solution used to reduce the acid attack on steel, while allowing it to attack iron oxides.

The hydrochloric acid used in the pickling solution has a 33% concentration and is yellow. The inhibitor and sodium sulfate originally dissolved in distilled water were added to it, and then distilled water was added up to 2 liters.

3.2.2.3 Rinsing

The rinsing stages after degreasing and pickling may consist of rinsing under running water or of immersion for a few seconds. In order to remove the chemical compounds present on the surface of the carbon steel samples after degreasing and pickling, the samples were immersed in distilled water [19].

3.2.2.4 Phosphating

The next step after the sample surfaces have been properly cleaned is phosphating. It consists of the formation of an insoluble phosphate film, resistant to corrosion, on the surface of the samples.

Phosphating solutions may be classified according to the nature of the metal ions that constitute the major component of the phosphating solution as follows: zinc, manganese or iron phosphate baths [41, 42]. The compositions of these solutions are chosen both depending on the material to be phosphated and depending on the expected properties.

In order to obtain a phosphate coat that will improve the corrosion resistance of carbon steel, as well as a coat that can be used as a substrate for future coatings, the solutions shown in Tables 3.9, 3.10 and 3.11 were used. The quantities of active substances used were calculated for 2 liters of phosphating solution, being supplemented with double-distilled water.

Table 3.9 Chemical composition of I phosphate solution.

Substance	Quantity
Sodium hydroxide (NaOH)	0.75 g
Sodium nitrite (NaNO$_2$)	0.45 g
Sodium tripolyphosphate (Na$_5$P$_3$O$_{10}$)	0.05 g
Phosphoric acid (H$_2$PO$_4$)	12.00 ml
Nitric acid (HNO$_3$)	6.00 ml
Zinc (Zn)	5.00 g

Table 3.9 Chemical composition of II phosphate solution.

Substance	Quantity
Sodium hydroxide (NaOH)	0.75 g
Sodium nitrite (NaNO$_2$)	0.45 g
Sodium tripolyphosphate (Na$_5$P$_3$O$_{10}$)	0.05 g
Phosphoric acid (H$_2$PO$_4$)	10.00 ml
Nitric acid (HNO$_3$)	4.00 ml
Zinc (Zn)	3.50 g
Iron (Fe)	0.038 g

Table 3.10 Chemical composition of III phosphate solution.

Substance	Quantity
Sodium hydroxide (NaOH)	0.75 g
Sodium nitrite ($NaNO_2$)	0.45 g
Sodium tripolyphosphate ($Na_5P_3O_{10}$)	0.05 g
Phosphoric acid (H_2PO_4)	7.00 ml
Nitric acid (HNO_3)	0.40 ml
Nickel (Ni)	0.03 g
Iron (Fe)	0.03 g
Manganese (Mn)	1.50 g

In Romania, the STAS 7969-85 standard lists the main categories of phosphating solutions, as well as the amounts and ratios to take into account for their development. Phosphate coats are deposited on the surface of the material by spraying or immersion processes, the most appropriate method being chosen depending on the size and shape of the surface to be phosphated, as well as on its subsequent use. Although the immersion process takes longer, due to the even coating of the material surface and taking into account the size, shape and manufacturing process of the carabiner, this method was thought to be the most suitable for phosphate deposition. The samples were immersed in the phosphating solutions at 95 °C for 30 minutes. The role of the active components added in the phosphating solutions is described below.

Phosphoric acid

All conventional phosphating baths include solutions based on phosphoric acid diluted by one or more metal ions of alkali or heavy metals. These baths contain free phosphoric acid, as well as primary phosphates of metal ions. When the carbon steel part is immersed in the phosphating solution, a reaction takes place which results in a significant change in the structure of the material, also called topochemical reaction, during which the dissolution of the base material structure begins due to the presence of free phosphoric acid in the bath.

In order to suppress hydrolysis and to keep the bath stable for the deposition of the phosphate layer, there must be a certain amount of free phosphoric acid in the bath. Therefore, depending on the temperature of the phosphate bath and the concentration of its constituents, the necessary content of free phosphoric acid is determined in order to maintain the equilibrium conditions. An orthophosphoric acid with 85% concentration was used to make the three phosphating solutions.

Nitric acid

Nitric acid is a colorless liquid, easily soluble in water, with the chemical formula HNO_3^- The acid used to make the phosphating solutions has a 65% concentration. It was added over phosphoric acid to accelerate the dissolution of metal ions added later.

Sodium hydroxide

0.75g of sodium hydroxide was added to 2l of phosphating solution to provide chemical stability to the phosphate coat, and to inhibit the reaction between free phosphoric acid and the surface of the material. Thus, trisodium phosphate and water were formed following the reaction between sodium hydroxide and phosphoric acid.

Sodium nitrite

Sodium nitrite is a salt of nitric acid, in the form of colorless water-soluble crystals, whose chemical formula is $NaNO_2$. The amount of 0.45g per 2 liters of solution was initially dissolved in 30ml of double-distilled water, then added to the phosphating solution to accelerate the formation of the phosphate coat on the steel surface. Moreover, sodium nitrite is also used as an oxidizing agent that reacts with hydrogen resulting from the reaction between phosphoric acid and metal ions, forming water, which prevents the formation of a passivating coat on the metal surface [43].

Sodium tripolyphosphate

Sodium tripolyphosphate is a white inorganic compound having the chemical formula $Na_5P_3O_{10}$. It has been added to phosphating solutions due to its role of activating the metal surface, as well as inhibiting corrosion, extending the life of the bath by reducing the consumption of chemicals. The amount of 0.05 g was originally dissolved in double-distilled water.

Zinc

Zinc is the most widely used anti-corrosion, silver-colored, metallic chemical element obtained from ores and compounds. It was used in the form of chips in both solutions. In both the first and the second phosphating solution, zinc was added to the concentrated solution of phosphoric acid and nitric acid. Following the addition of zinc, an exothermic reaction took place as a result of which zinc phosphate was formed, as shown in Fig. 3.7.

Figure 3.7 Zinc dissolution: a) Zinc chips; b) Zinc dissolution reaction.

Manganese

Due to its deoxidation properties, manganese is an important element in the manufacture of steels. The third phosphating solution was made by dissolving 1.5 g of manganese chips in the nitric acid and phosphoric acid solution. An exothermic reaction occurred, which led to the formation of manganese phosphate (Fig. 3.8).

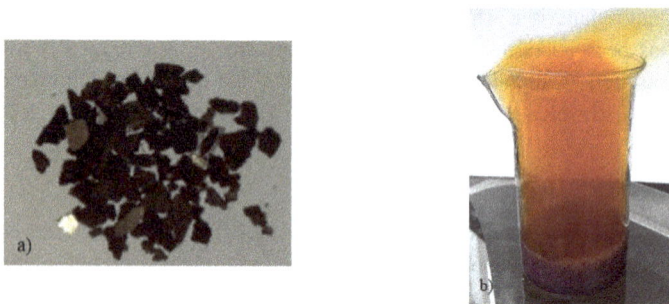

Figure 3.8 Manganese dissolution: a) Manganese chips; b) Manganese dissolution reaction.

Iron

As a result of the reactions between phosphoric acid and zinc or manganese ions in the phosphating solution, hydrogen is formed. Iron powder was added to the second and third solutions in order to prevent the negative effects on the deposition, by the decreasing hydrogen content.

Nickel

Nickel is a white-gray metal. It was added as chips to the third phosphating solution in order to improve the surface texture, providing better adhesion to subsequent coatings, as well as higher corrosion resistance.

3.2.2.5 Sample rinsing after phosphating

After phosphating, the samples are rinsed with distilled water in order to remove soluble active salts, which may lead to pockets occurring under the following coat of paint.

3.2.2.6 Drying

After rinsing, before the phosphating process is completed, the samples must be dried. The available drying methods are evaporation, forced drying by air currents or sample heating.

3.2.3   Paint coating technology

In order to improve their shock resistance properties, a coat of paint with high elasticity properties was sprayed on the surface of the carbon steel samples and on the surface of the phosphate samples [44].

A prior coat deposited by conversion is essential for the durability of painted metal products. Due to the porosity of the phosphate coat, these coatings improve the adherence of the coat of paint deposited later. The technology of elastomer-based paint coating is shown in Fig. 3.9.

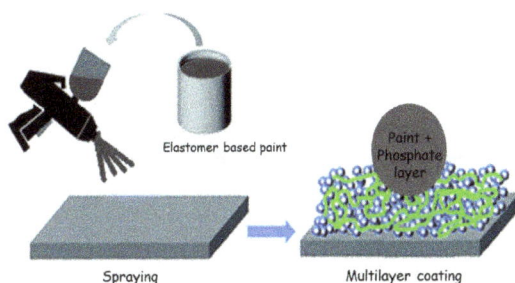

Figure 3.9 Paint coating technology.

Considering that zinc phosphate coatings provide good durability of painted parts in corrosive environments, the use of zinc phosphate obtained from the first phosphating solution was chosen for the mechanical characterization of painted phosphate steel [.

The paint that we used, KS-1000, Car System, was purchased at the store. This is an elastomer-based product used to paint cars, which provides protection against external factors, such as impact with stones or other objects, due to its elasticity.

3.3   Methods of analysis and equipment used in experimental research

The resulting zinc and manganese phosphate coats, as well as the paint coat subsequently deposited were analyzed by structural, surface, thermal and mechanical characterization, and their corrosion resistance was determined. The used samples were cut with an EDM cutting device, while observing the specific dimensions of each type of experimental research.

3.3.1   *Energy dispersive X-ray spectroscopy (EDX)*

In order to characterize the zinc phosphate and manganese phosphate coats deposited on the surface of carbon steel samples, one needs to know their chemical composition, as well as the concentration of components or impurities in the coat. Energy dispersive X-ray spectroscopy (EDX) is a technique for identifying and quantifying chemical compositions used in conjunction with the scanning electron microscope (SEM).

This method, shown in Fig. 3.10, detects the wavelengths specific to X-rays, characteristic of each atomic structure, emitted by the sample during its bombardment with an electron beam, in order to characterize the elemental composition of the tested surface. It may be determined in specific spots or in well-defined areas on the surface of the sample [47].

In order to determine the chemical composition of the phosphate coats deposited on the surface of carbon steel samples, we used an EDAX Bruker detector attached to a Vega Tescan LMH II scanning electron microscope owned by the Materials Science Department of the Faculty of Materials Science and Engineering, 'Gheorghe Asachi' Technical University of Iaşi.

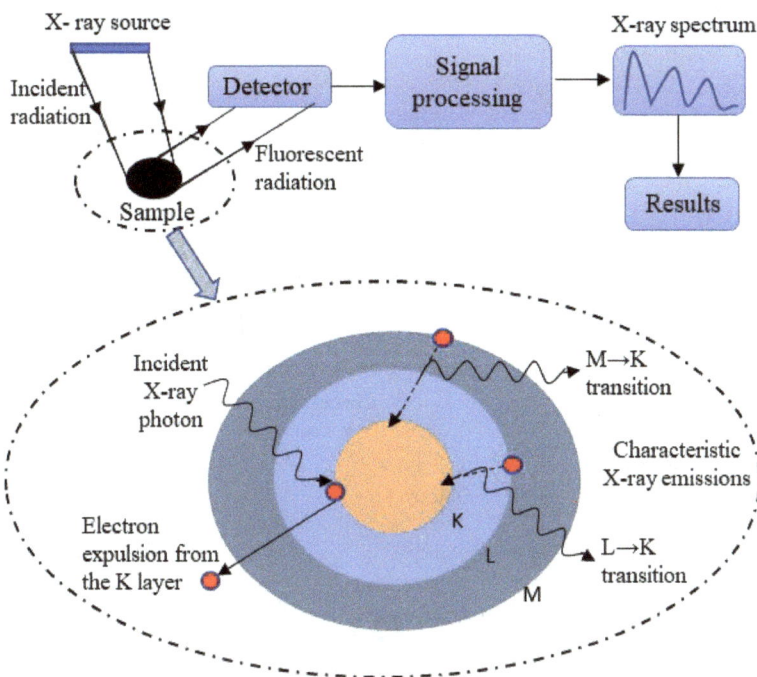

Figure 3.10 Operating scheme of the EDAX system [46].

3.3.2   Optical microscopy (OM)

Optical microscopy analysis provides images of the morphology and size of the crystals on the surface of the phosphate coat, with magnification powers ranging between 50X and 1000X. Optical microscopy is the method of analyzing the structure of materials, in which light is directed vertically through the lens of the microscope and reflected back through the lens to an eyepiece, viewing screen or camera.

The magnification of the image of the samples is obtained by light reflection through a combination of lenses and eyepieces. The minimum resolution of the features is about 0.2 μm. However, lower resolutions - about 0.05 μm - can be detected by improving the image contrast with polarized light, interference contrast and dark field illumination. The principle of operation of such a microscope is described in Fig. 3.11.

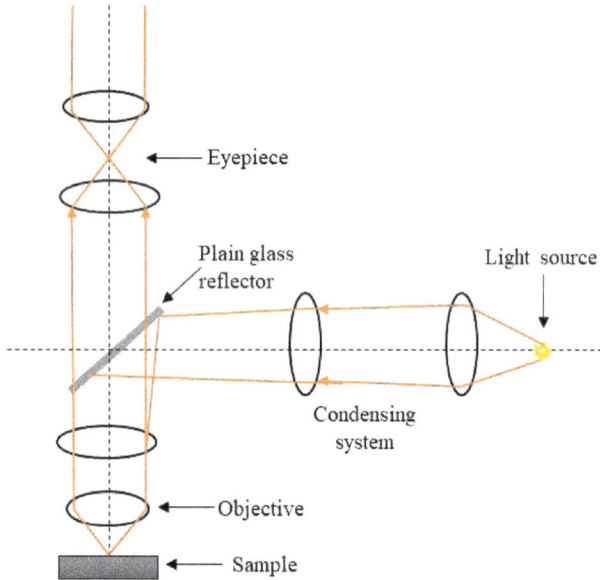

Figure 3.11 Operation principle of a metallographic optical microscope [48].

The microstructure specific to the crystals formed on the surface of carbon steel after phosphating was determined by an AxioCam 105 Zeiss Imager Axio a1M metallographic optical microscope equipped with a digital USB camera and AxioVision SE64 software. This device is owned by the Interdisciplinary Laboratory for Scientific Research and Conservation of Cultural Heritage Assets within 'Alexandru Ioan Cuza' University of Iaşi.

3.3.3   Scanning electron microscopy (SEM)

Scanning electron microscopy is a method for high-resolution imaging of surfaces. It uses electrons for imaging, just as an optical microscope uses visible light. Compared to optical microscopy, the image can be magnified up to 50000X, while the depth of field is up to 100X larger. In the case of scanning electron microscopy, the surface of the sample is scanned using an electron beam, and the interaction between it and the material results in the emission of electrons and photons as electrons enter the surface of the sample. The emitted particles are collected using a detector to provide information about the surface.

The surface image is the result of the collision of the electron beam with the surface of the tested sample. The principle of operation of a scanning electron microscope is shown in Fig. 3.12.

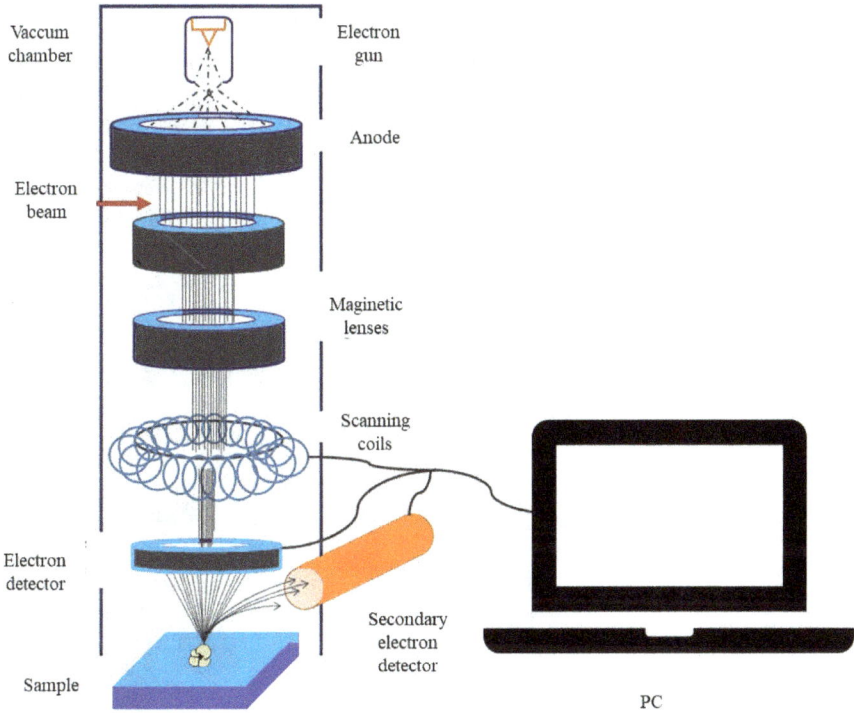

Figure 3.12 Operation principle of an electron scanning microscope [49].

Due to the use of carabiners, the coating process must be subjected to thorough quality checks to ensure the performance of the coating. Verification of the quality of coatings by conversion and analysis of crystal morphology were done using a Vega Tescan LMH II scanning electron microscope owned by the Department of Materials Science of the Faculty of Materials Science and Engineering, 'Gheorghe Asachi' Technical University of Iaşi.

3.3.4   Fourier transform infrared spectroscopy (FTIR)

Fourier Transform Infrared Spectroscopy (FTIR) is an analytical technique used to identify inorganic and organic materials, by monitoring variations in characteristic structural groups and vibrations, which provides an evaluation of the tested materials.

FTIR spectroscopy has many advantages when used for the chemical analysis of phosphate coats. The resulting spectrogram provides useful information about the type of compound existing on the sample surface, as well as about the composition of the phases. This device uses an interferometer to refract the wavelength of a source that emits infrared radiation. The intensity of the reflected light is measured with a detector and quantified according to its wavelength. The result obtained, an interferogram, is to be analyzed by a software that uses Fourier transforms to obtain an infrared spectrum. Its operating principle is described in Fig. 3.13.

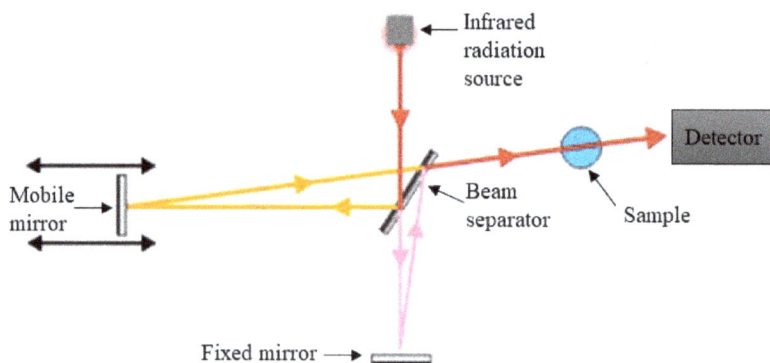

Figure 3.13 Operation principle of an FTIR spectrometer [50].

The determination of the chemical structure of the phosphate coats deposited on the surface of carbon steel samples after phosphating was done by attenuated total reflection transmission (ATR) using a Bruker Hyperion 1000 FTIR spectrometer, equipped with a 15X lens, the resulting spectra being analyzed by the OPUS 65 Bruker software. The equipment is owned by the Interdisciplinary Laboratory for Scientific Research and Conservation of Cultural Heritage Assets within 'Alexandru Ioan Cuza' University of Iaşi.

3.3.5   Determination of mechanical properties by scratch and microindentation tests

The scratch test is a quick and easy way to characterize coatings. This test determined the adherence of the zinc phosphate coat and of the paint coat subsequently deposited, as well as their coefficient of friction.

In this method, the sample is subjected to a continuously increasing force, while being moved at constant speed. The thickness of the coat, the mechanical properties of the substrate and the test conditions must be taken into account, as they may influence the test results [45, 51].

The microindentation test is used to determine the mechanical properties of metals, due to the ease and speed with which it can be carried out [52]. It was performed to evaluate changes in the mechanical properties of the deposited coats, such as hardness and modulus of elasticity. The principle of performing scratch and microindentation tests is described in Fig. 3.14.

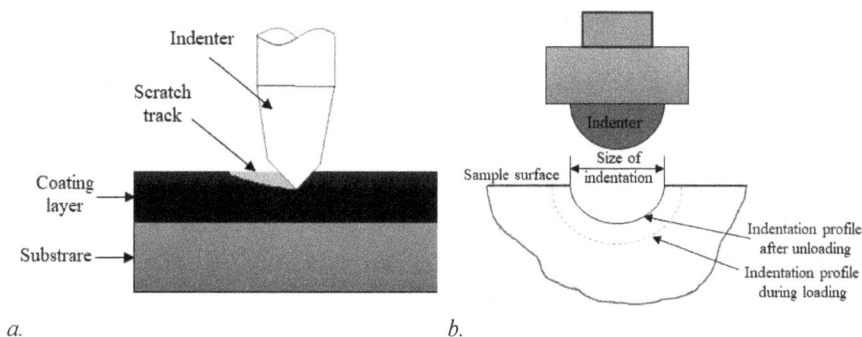

Figure 3.14 The principle of performing the tests: a) scratching; b) microindentation [53].

The mechanical properties of the deposited phosphate and paint coats were determined using a CETR UMT-2 universal micro-tribometer owned by the Tribology Laboratory of the Faculty of Mechanics, 'Gheorghe Asachi' Technical University of Iaşi.

3.3.6   Impact test

The Charpy impact test is a standardized test, which helps to determine the amount of energy absorbed by a material during its breaking due to shock [54]. This absorbed energy is a measure of the strength of the material and acts as a tool for studying the ductile-brittle transition as a function of temperature [55].

The device consists of a pendulum with a known mass and length, which is dropped from a standard height (Fig. 3.15). The energy transferred to the material can be inferred by comparing the difference in hammer height between the starting position (before breaking) and the final position (after breaking) [44].

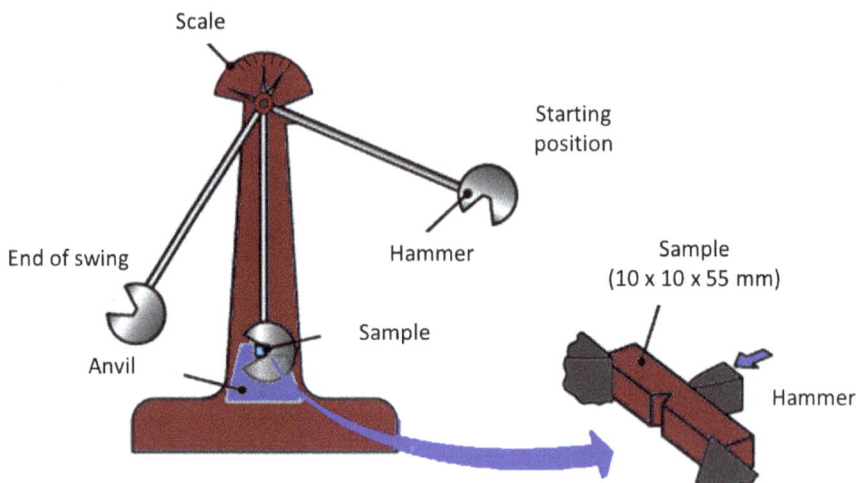

Figure 3.15 Charpy Impact Test [56].

The impact resistance of phosphated and painted samples was tested according to the EN 10045-1:1990 standard, using a pendulum for bending by shock - Charpy Hammer owned by the Laboratory of Plastic Deformation within the Department of Technologies and Equipment for Materials Processing, Faculty of Materials Science and Engineering, 'Gheorghe Asachi' Technical University of Iași.

3.3.7   Corrosion resistance

Corrosion is the physical-chemical, spontaneous and irreversible destruction of metals or alloys under the chemical, electrochemical or biological action of the environment [57].

Corrosion processes depend on the nature of the metal and the corrosive environment, on pressure and temperature, and on the static or dynamic conditions of the corrosive environment [58].

Most natural corrosion processes take place through electrochemical reactions that occur in humid environments. For this reason, electrochemical study methods are a fast and efficient means of obtaining information on the thermodynamic probability of corrosion of a metal immersed in a liquid, instantaneous corrosion rate (corrosion rate at simple immersion of the metal in the corrosion environment), the type of corrosion, as well as the factors that may influence corrosion (temperature, pH, corrosion accelerators or inhibitors) [59].

The corrosion behavior of C45 steel samples, C45 phosphate steel with three types of phosphating solutions with different compositions, as well as C45 steel samples phosphated with the first solution and immersed in MoS_2-based lubricant or painted C45 steel samples was analyzed. The following notations have been used to simplify writing:

$C45$ – C45 steel sample;

$I - Zn-$ C45 sample phosphated in the first zinc solution;

$II - Zn/Fe$ – C45 sample phosphated in the second zinc and iron solution;

$III - Mn$ - C45 sample phosphated in the third manganese solution;

$OFU-$ C45 sample phosphated in the first solution and immersed in MoS_2-based lubricant;

$OFV-$ C45 sample phosphated in the first solution and painted.

The liquids used in this thesis as corrosion media were marked:

APL – Rain water; pH=6.5

AMN – Black Sea water; pH=6.15

SSI – Fire extinction solution; pH=6.41

The pH of these solutions was determined with a RADELKIS OP-264/1 pH-meter owned by 'Cristofor Simionescu' Faculty of Chemical Engineering and Environmental Protection, 'Gheorghe Asachi' Technical University of Iaşi.

Electrochemical corrosion studies and corroded surfaces characterization were performed using potentiostats: the PGP201 potentiostat was used for linear and cyclic polarization measurements, data processing was done with VoltaMaster 4 software, while the PGZ301 potentiostat was used to perform electrochemical impedance spectroscopy measurements, the data being processed using the ZSimpWin program, in which the spectra are interpreted by the fitting process developed by Boukamp - by the least squares method. In order to process with this program data purchased with the VoltaMaster 4 program, they were converted using the EIS file converter program [41].

A C145/170 three-electrode corrosion cell was used for both potentiodynamic measurements and electrochemical impedance spectroscopy (SIE) determinations. This is a glass cell where the liquid corrosion medium can flow freely. The samples (flat discs 10 mm in diameter) were fastened in the working electrode by means of Teflon washers, which allowed the creation of flat circular surfaces of up to 0.8 cm2. In the tested samples, the surface exposed to the corrosion environment was $S = 0.503$ cm^2. A flat platinum electrode ($S = 0.8$ cm^2) was used as an auxiliary electrode, and a saturated calomel electrode as a reference electrode. All potentials were measured in relation to this electrode, but, for simplicity of writing, the tables and the text do not mention it. The solutions used were naturally aerated [17].

The working conditions used in the measurements were as follows:

- linear anodic polarization: potential range (-200) ÷ (+300) mV compared to the open circuit potential, potential scanning speed: dE / dt = 0.5 mV/s;

- cyclic polarization: potential range (-800) ÷ (+2000) mV, potential scanning speed = 10 mV/s;

- SIE measurements: working potential = open circuit potential, frequency range = 104 ÷ 210 Hz, current amplitude = 10 mV.

The non-phosphated samples were sanded on metallographic paper up to 2500 grains/ mm^2, degreased and rinsed in distilled water. The phosphated samples, the ones impregnated in lubricant, as well as the painted ones were rinsed in distilled water.

3.4 Equipments

Due to the various fields in which they are used, carabiners come into contact with various corrosive substances. Therefore, in order to be safe, they need to have good corrosion resistance properties [60,61]. The deposition of coats of insoluble crystalline phosphate on the surface of the carbon steel used to manufacture carabiners is a simple and inexpensive method to improve its corrosion resistance properties.

Due to the porosity of the phosphate coats, they can be successfully used as a base for other types of coatings. Improving the impact resistance of carbon steel can be achieved by spraying a coat of elastomer-based paint on its phosphated surface.

In order to demonstrate that the carbon steel used to manufacture carabiners was improved by depositing phosphate and paint coats on its surface, a series of laboratory researches were performed, using high-performance equipment, which helped to characterize the deposited coats from the chemical, structural, physical and mechanical points of view.

The equipment used in the laboratory research, as well as the type of samples used are shown in Table 3.10.

Table 3.10 Laboratory research and equipment used to carry out the research.

Test	Sample	Equipment used for samples characterization
X-ray spectroscopy by energy dispersion (EDX)		Electron scanning microscope, model Vega Tescan LMH II equipped with EDAX Bruker detector
Optical microscopy		Zeiss Imager Axio a1M
Scanning electron microscopy		Electron scanning microscope, model Vega Tescan LMH II

Fourier Transform Infrared Spectroscopy (FTIR)		Spectometer FTIR Bruker Hyperion 1000
Scratch and microindentation tests		Universal Micro-tribometer CETR UMT-2
Impact test		Charpy hammer
Corrosion resistance		Potentiostat and three-electrode cell

The experimental research carried out with this equipment in our research allowed the characterization of the structural, mechanical and chemical properties of the phosphate and paint coats deposited on the surface of carbon steel samples in order to improve the properties of the carabiner.

References

[1] D.P. Burduhos-Nergis, C. Baciu, P. Vizureanu, N.M. Lohan, C. Bejinariu, Materials types and selection for carabiners manufacturing: a review, Euroinvent ICIR IOP Conference Series: Materials Science and Engineering 572 (2019) 012027. https://doi.org/10.1088/1757-899X/572/1/012027

[2] D.P. Burduhos Nergiş, C. Nejneru, D.C. Achiţei, N. Cimpoieşu, C. Bejinariu, Structural Analysis of Carabiners Materials Used at Personal Protective Equipments. In Proceedings of the IOP Conference Series: Materials Science and Engineering; Institute of Physics Publishing, 374, 2018. https://doi.org/10.1088/1757-899X/374/1/012040

[3] M.H. Hafiz, J.S. Kashan, A.S. Kani, Effect of Zinc Phosphating on Corrosion Control for Carbon Steel Sheets. Eng. & Technology, 26(5) (2008) 501-511.

[4] A.V. Sandu, C. Bejinariu, G. Nemtoi, I.G. Sandu, P. Vizureanu, I. Ionita, C. Baciu, New anticorrosion layers obtained by chemical phosphatation, REVISTA DE CHIMIE, 64(8) (2013) 825-827.

[5] A.V. Sandu, C. Coddet, C. Bejinariu, Study on the chemical deposition on steeel of zinc phosphate with other metallic cations and hexamethilen tetramine. I. Preparation and structural and chemical characterization, Journal of Optoelectronics and Advanced Materials, 14(7-8) (2012) 699 - 703.

[6] A.B. Forero, M.G. Nunez Milagros, I.S. Bott, Analysis of the corrosion scales formed on API 5L X70 and X80 steel pipe in the presence of CO_2. Mat. Res., 17(2) (2013) 461-471. https://doi.org/10.1590/S1516-14392013005000182

[7] H. Jafari, I. Danaee, H. Eskandari, M. Rashvandavei, Electrochemical and quantum chemical studies of N,N'-bis(4-hydroxybenzaldehyde)-2,2-dimethylpropandiimine Schiff base as corrosion inhibitor for low carbon steel in HCl solution. J Environ Sci Health a Tox Hazard Subst Environ Eng., 48(13) (2013) 1628-1641. https://doi.org/10.1080/10934529.2013.815094

[8] A.V. Sandu, C. Coddet, C. Bejinariu, A Comparative Study on Surface Structure of Thin Zinc Phosphates Layers Obtained Using Different Deposition Procedures on Steel, REVISTA DE CHIMIE, 63(4) (2012) 401-406.

[9]   C.R. Tomachuk, C.I. Elsner, A.R. Di Sarli, Electrochemical characterization of chromate free conversion coatings on electrogalvanized steel. Mat. Res., 17(1) (2013) 61-68. https://doi.org/10.1590/S1516-14392013005000179

[10]   C.A. Cunha, N.B. Lima, J.R. Martinelli, A.H.A. Bressiani, A.G.F. Padial, L.V. Ramanathan, Microstructure and mechanical properties of thermal sprayed nanostructured Cr3C2–Ni20Cr coatings. Mat. Res., 11(2) (2008) 137-143. https://doi.org/10.1590/S1516-14392008000200005

[11]   E.S. Bacaita, C. Bejinariu, B. Zoltan, C. Peptu, G. Andrei, M. Popa, D. Magop M. Agop, Nonlinearities in Drug Release Process from Polymeric Microparticles: Long-Time-Scale Behaviour, J. Appl. Math. 653720, 2012. https://doi.org/10.1155/2012/653720

[12]   K. Abdalla, A. Rahmat, A. Azizan, Influence of activation treatment with nickel acetate on the zinc phosphate coating formation and corrosion resistance. Materials and Corrosion, 65 (2014) 977-981. https://doi.org/10.1002/maco.201307009

[13]   C. Bejinariu, D.P. Burduhos-Nergiş, N. Cimpoeşu, M.A. Bernevig-Sava, Ş.L. Toma, D.C. Darabont, C. Baciu, Study on the anticorrosive phosphated steel carabiners used at personal protective equipment. Quality - Access to Success, 20 (2019) 71–76.

[14]   C. Nejneru, M.C. Perju, D.D. Burduhos Nergis, A.V. Sandu, C. Bejinariu, Galvanic Corrosion Behaviour of Phosphate Nodular Cast Iron in Different Types of Residual Waters and Couplings, REVISTA DE CHIMIE, 70(10) (2019) 3597-3602. https://doi.org/10.37358/RC.19.10.7604

[15]   A.V. Sandu, A. Ciomaga, G. Nemtoi, C. Bejinariu, I. Sandu, Study on the chemical deposition on steeel of zinc phosphate with other metallic cations and hexamethilen tetramine. II. Evaluation of corrosion resistance, Journal of Optoelectronics and Advanced Materials, 14(7-8) (2012) 704-708.

[16]   A.V. Sandu, C. Bejinariu, A. Predescu, I.G. Sandu, C. Baciu, I. Sandu, New mechanisms for chemical phosphatation of iron objects, RECENT PATENT ON CORROSION SCIENCE, (ISSN 1877-6108), Bentham Science Publishers, 1(1) (2011) 33-37. https://doi.org/10.2174/2210683911101010033

[17]   D.P. Burduhos-Nergis, P. Vizureanu, A.V. Sandu, C. Bejinariu, Evaluation of the Corrosion Resistance of Phosphate Coatings Deposited on the Surface of the Carbon Steel Used for Carabiners Manufacturing, Applied Sciences 10(8) (2020) 2753. https://doi.org/10.3390/app10082753

[18]   A.V. Sandu, C. Bejinariu, I.G. Sandu, M.M.A.B. Abdullah, Modern Technologies

of Thin Films Deposition. Chemical Phosphatation, Material Research Forum, USA (ISBN 978-1-945291-90-6) (2018) 149.

[19] D.P. Burduhos-Nergis, C. Bejinariu, S.L. Toma, A.C. Tugui, E.R. Baciu, Carbon steel carabiners improvements for use in potentially explosive atmospheres. MATEC Web of Conferences, 305, 00015, 2020. https://doi.org/10.1051/matecconf/202030500015

[20] C. Bejinariu, I. Sandu, V. Vasilache, I.G. Sandu, M.G. Bejinariu, A.V. Sandu, M. Sohaciu, V. Vasilache, Process for the micro-crystalline phosphate-coating of iron-based metal pieces, Patent RO125457-B1/2014.

[21] E. Thorpe, A Dictionary of Apllied Chemistry, Vol. V. Ed. Longmans, Green, and Co., London, England, 1913. Information on: https://archive.org/details/dictionary ofappl05thorrich/page/n5 (Accessed: August 26, 2019).

[22] Safety Data Sheet Sodium hydroxide, solid SDS for Lye (Sodium Hydroxide) printed from Section 1-Identification Section 2-Hazards Identification Emergency Overview Section 4-First Aid Measures (2019-08-12). Information on: http://www.certified-lye.com/SDS-Lye.pdf (Accessed: August 12, 2019).

[23] W.N. Hussein, S.S. Bahar, N.M. Abdilrida, Optimization of Using NaOH in Industrial Cleaning Systems. Journal of Babylon University Engineering Sciences, 22(1) (2014) 78-85.

[24] Soda Ash Statistics and Information (2019 08 12). Information on: https://www.usgs.gov/centers/nmic/soda ash statistics and information?qt science_support_page_related_con=0#qt science_support_page_related_con (Accessed: August 12, 2019).

[25] Sodium metasilicate 6834 92 0 (2019 08 12). Information on: https://www.chemicalbook.com/ChemicalProductProperty_EN_CB2199386.htm (Accessed: December 12, 2019).

[26] K. Schrodter, G. Bettermann, T. Staffel, F. Wahl, T. Klein, T. Hofmann, Ullmann's Encyclopedia of Industrial Chemistry, Phosphoric Acid and Phosphates, Wiley-VCH Verlag GmbH & Co, 2008. https://doi.org/10.1002/14356007.a19_465

[27] Trisodium phosphate| 7601 54 9 (2019 08 12). Information on: https://www.chemicalbook.com/ChemicalProductProperty_EN_cb4364607.htm (Accessed: August 12, 2019).

[28] SILMACO. Sodium Metasilicate: Green and Efficient. Technical Brochure, Version 2 from 28.11.2016.

[29] V. Nedeff, C. Bejenariu, G. Lazar, M. Agop, Generalized lift force for complex

fluid. Powder Technol. 235 (2013) 685–695.
https://doi.org/10.1016/j.powtec.2012.11.027

[30] P.-E. Nica, M. Agop, S. Gurlui, C. Bejinariu, C. Focsa, Characterization of Aluminum Laser Produced Plasma by Target Current Measurements. Jpn. J. Appl. Phys. 51 (2012) 106102. https://doi.org/10.1143/JJAP.51.106102

[31] The Chemistry of Cleaning Essential Industries (2019 08 12). Information on: https://www.essind.com/chemistry-of-cleaning/ (Accessed: August 12, 2019).

[32] L.M. Fernandez Diaz, Study of the inhibitors in the acidic attack on steel surfaces during the elimination of oxide scales. Dissertation Thesis, Hamburg, Germania, 2008.

[33] E. Bayol, K. Kayakirilmaz, M. Erbil, The inhibitive effect of hexamethylenetetramine on the acid corrosion of steel. Materials Chemistry and Physics, 104 (2007) 74-82. https://doi.org/10.1016/j.matchemphys.2007.02.073

[34] M. Kumari, Use of hexamine as corrosion inhibitor for carbon steel in hydrochloric acid. International Journal of Advanced Educational Researc, 2(6) (2017) 224-235.

[35] R.C. Wells, Sodium sulphate: its sources and uses. Bulletin 717, Government Printing Office, Washington, United States of America 1923.

[36] N. Ipek, N. Lior, A. Eklund, Improvement of the electrolytic metal pickling process by inter-electrode insulation. Ironmaking and Steelmaking, 32(1) (2005) 87-96. https://doi.org/10.1179/174328105X23996

[37] OxyChem, Hydrochloric Acid Handbook, Occidental Chemical Corporation 2018.

[38] C.J. Brown. Productivity improvements through recovery of pickle liquors with the APU process. Iron and Steel Engineer Journal, 55-60 1990.

[39] M. Maanonen, Steel Pickling in Challenging Conditions. Bachelor Thesis, Helsinki, Finlanda, 2014.

[40] T.J. Fox, C.D. Randall, D.H. Gross, Steel Pickling: A Profile. Draft Report, EPA Contract Number 68-D1-0143 & RTI Project Number 35U-5681-58 DR, 1993.

[41] D.P. Burduhos-Nergis, P. Vizureanu, A.V. Sandu, C. Bejinariu, Phosphate Surface Treatment for Improving the Corrosion Resistance of the C45 Carbon Steel Used in Carabiners Manufacturing, Materials 13(15) (2020) 3410. https://doi.org/10.3390/ma13153410

[42] D.P. Burduhos-Nergis, C. Nejneru, D.D Burduhos-Nergis, C. Savin, A.V Sandu, S.L Toma, C. Bejinariu, The Galvanic Corrosion Behavior of Phosphated Carbon Steel Used at Carabiners Manufacturing, Revista de chimie 70(1) (2019) 215-219.

https://doi.org/10.37358/RC.19.1.6885

[43] J.L. Stauffer, Finishing Systems Design and Implementation: A Guide for Product Parameters, Coatings, Process, and Equipment. Society of Manufacturing Engineers, ISBN: 978-0872634343, Michigan, United States of America, 1993.

[44] D.P. Burduhos-Nergis, A.V. Sandu, D.D. Burduhos-Nergis, D.C. Darabont, R.-I. Comaneci, C. Bejinariu, Shock Resistance Improvement of Carbon Steel Carabiners Used at PPE, MATEC Web Conf. 290 (2019) 12004. https://doi.org/10.1051/matecconf/201929012004

[45] D.P. Burduhos Nergis, N. Cimpoesu, P. Vizureanu, C. Baciu, C. Bejinariu, Tribological characterization of phosphate conversion coating and rubber paint coating deposited on carbon steel carabiners surfaces, Materials today: proceedings 19 (2019) 969-978. https://doi.org/10.1016/j.matpr.2019.08.009

[46] S. Qutaishat, Silicon Drift Detectors for Synchrotron Energy Dispersive X-Ray Fluorescence Spectroscopy (SDD for EDXRF), Conference Paper - Presentation, The ninth SeSAMe users'meeting, Amman, Jordan, 2011.

[47] A.V. Sandu, A. Ciomaga, G. Nemtoi, C. Bejinariu, I. Sandu, SEM-EDX and microFTIR studies on evaluation of protection capacity of some thin phosphate layers, Microscopy Research and Technique, 75(12) (2012) 1711-1716. https://doi.org/10.1002/jemt.22120

[48] B. Biswas, Growth Defects in CrN/NbN Coatings Deposited by HIPIMS/UBM technique. pHD Thesis, Sheffield Hallam University, Olanda, 2017.

[49] Spotlight on Science: Scanning Electron Microscopy (2019 08 07). Information on: https://geologelizabeth.wordpress.com/2014/03/26/spotlight on science scanning electron-microscopy/ (Accessed: August 07, 2019).

[50] What is FTIR Spectroscopy? (2019 08 07). Information on: http://www.tech-faq.com/what-is-ftir-spectroscopy.html (Accessed: August 13, 2019).

[51] R.S.R. Kalidindi, R. Subasri, Sol-gel nanocomposite hard coatings, in Anti-Abrasive Nanocoatings. Anti-Abrasive Nanocoatings. Woodhead Publishing, 105-136, ISBN: 978-0-85709-211-3, Cambridge, United Kingdom, 2015. https://doi.org/10.1016/B978-0-85709-211-3.00005-4

[52] J. Zhang, H. Zhang, J. Wu, Fuel Cell Degradation and Failure Analysis Pem Fuel Cell Testing and Diagnosis. Elsevier, 283-335, ISBN: 9780444536884, Germany, 2013. https://doi.org/10.1016/B978-0-444-53688-4.00011-5

[53] D. Tzetzis, K. Tsongas, G. Mansow, Determination of the Mechanical Properties of Epoxy Silica Nanocomposites through FEA-Supported Evaluation of Ball

Indentation Test Results. Materials Research, 20(6) (2017) 1571-1578. https://doi.org/10.1590/1980-5373-mr-2017-0454

[54]   N. Saba, M. Jawaid, M.T.H. Sultan, An overview of mechanical and physical testing of composite materials. Mechanical and Physical Testing of Biocomposites, Fibre-Reinforced Composites and Hybrid Composites. 1-12, 2019. https://doi.org/10.1016/B978-0-08-102292-4.00001-1

[55]   C. Chandrasekaran, Testing of Rubber Lining, Anticorrosive Rubber Lining, Plastic Design Library. William Andrew, 165-172, ISBN: 9780323443715, New York, United States of America, 2017. https://doi.org/10.1016/B978-0-323-44371-5.00019-0

[56]   Green Mechanic: Difference Between Izod and Charpy Test (2019 08 07). Information on: https://www.green-mechanic.com/2014/04/difference-between-izod-and-charpy.html (Accessed: August 07, 2019).

[57]   C Bejinariu, D.P. Burduhos-Nergis, N. Cimpoesu, M.A. Bernevig-Sava, S.L. Toma, D.C. Darabont, C. Baciu, Study on the anticorrosive phosphated steel carabiners used at personal protective equipment, Quality-Access to Success 20(1) (2019) 71-76.

[58]   D.P. Burduhos-Nergis, C. Nejneru, R. Cimpoesu, A.M. Cazac, C. Baciu, D.C. Darabont, C. Bejinariu, Analysis of Chemically Deposited Phosphate Layer on the Carabiners Steel Surface Used at Personal Protective Equipments, Quality-Access to Success 20(1) (2019) 77-82.

[59]   P. Lazar, C. Bejinariu, A.V. Sandu, A.M. Cazac, O. Corbu, M.C. Perju, I.G. Sandu, Corrosion Evaluation of Some Phosphated Thin Layers on Reinforcing Steel, IOP Conference Series: Materials Science and Engineering, 209(1) (2017) 012025. https://doi.org/10.1088/1757-899X/209/1/012025

[60]   C. Bejinariu, D.-C. Darabont, E.-R. Baciu, I.-S. Georgescu, M.-A. Bernevig-Sava, C. Baciu, Considerations on Applying the Method for Assessing the Level of Safety at Work. Sustainability 9 (2017) 1263. https://doi.org/10.3390/su9071263

[61]   C. Bejinariu, D.C. Darabont, E.R. Baciu, I. Ionita, M.A.B. Sava, C. Baciu, Considerations on the Method for Self Assessment of Safety at Work. Environ. Eng. Manag. J. 16 (2017) 1395–1400. https://doi.org/10.30638/eemj.2017.151

CHAPTER 4

Structural Characterization of Phosphate Coats

4.1 Determination of the chemical composition of phosphate coats

The chemical composition of phosphate coats needs to be determined for their structural characterization [1]. Energy dispersive X-ray spectroscopy (EDX) was used to perform the elemental analysis.

Elemental analysis is the first laboratory study conducted in this research. It aims to determine the chemical composition of phosphate coats, and to check the evenness of the deposited coats.

Five EDX determinations were performed on the surface of each deposited coat in different areas of the sample, in order to determine its chemical composition. The elemental analysis was performed using the EDAX Bruker detector attached to the Vega Tescan LMH II scanning microscope.

The samples used to determine the chemical composition of the phosphate coats by energy dispersion X-ray spectroscopy are cylindrical and 10 mm in diameter and 3 mm in height, as can be seen in Fig. 4.1.

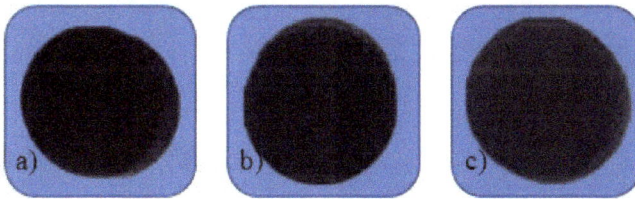

Figure 4.1 EDX sample used: a) I − Zn; b) II − Zn/Fe; c) III − Mn.

The test findings about the chemical composition of phosphate coats revealed the main elements identified in the coat.

Table 4.1 shows the mean of the chemical compositions of the phosphate coat deposited on carbon steel.

Table 4.1 Chemical composition of deposited phosphate layers.

Chemical composition	*I − Zn [%]*	*II − Zn/Fe [%]*	*III − Mn [%]*
P	12.46	13.72	1.85
Fe	15.16	20.78	79.63
O	36.54	37.70	12.24
Zn	35.83	27.77	-
Ni	-	-	3.82
Mn	-	-	2.72

Figs 4.2, 4.3 and 4.4 show the EDX spectra characteristic of the resulting phosphate coats, as well as their EDX microstructure.

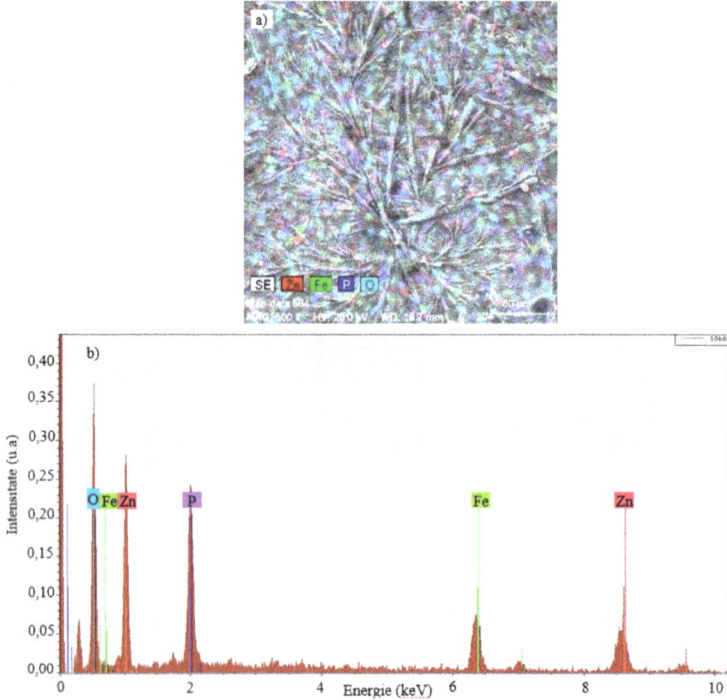

Figure 4.2 EDX analysis of the I − Zn sample: a) elementary chemical distribution; b) EDX spectre.

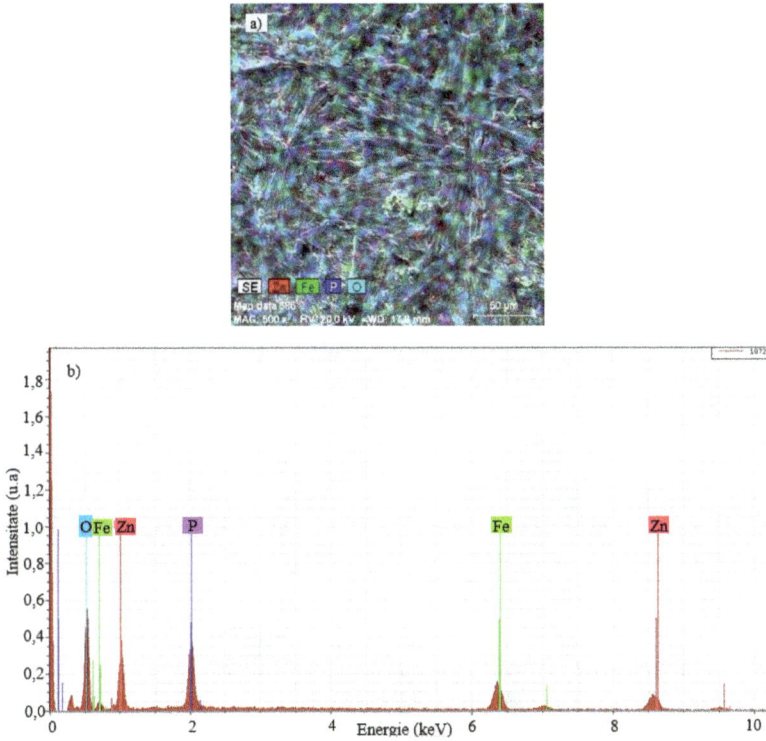

Figure 4.3 EDX analysis of the $II - Zn/Fe$ sample: a) elementary chemical distribution; b) EDX spectre.

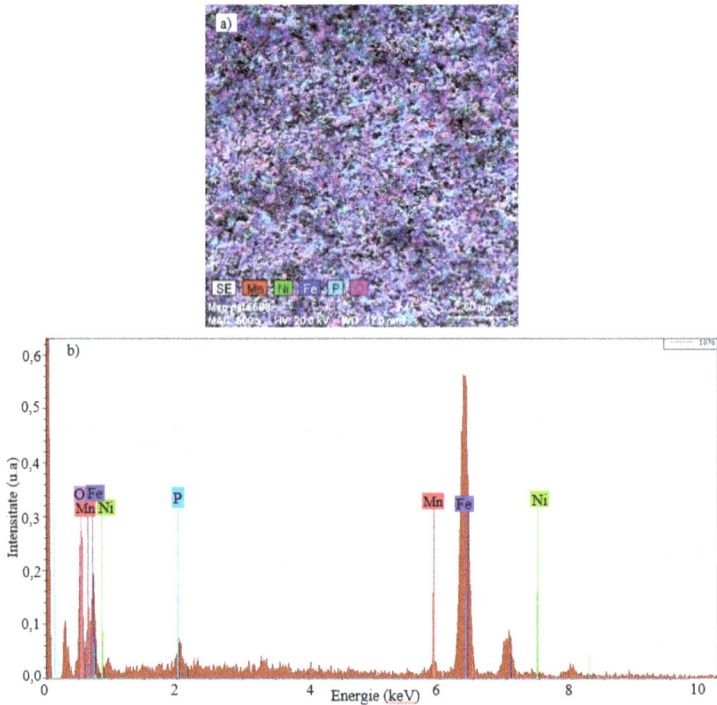

Figure 4.4 EDX analysis of the III − Mn sample:a) elementary chemical distribution; b) EDX spectre.

The atomic percentage of the constituents is revealed by the EDX spectra. The results confirm the presence of P, Fe and Zn/Mn, in the areas where there is coverage and no other element is present. In the II − Zn/Fe sample, iron may occur due to the formation of the crystalline phosphophyllite structure or from the contents of the steel substrate which may interfere due to X-ray penetration to the interface with the substrate [2]. The III − Mn sample results indicate the presence of nickel particles on the metal surface, particles that accelerated the formation of the coating.

4.2 Structural characterization of phosphate layers by light microscopy and scanning electron microscopy

Light microscopy and scanning electron microscopy are used to examine the structure and evenness of the crystals. The surface morphology of the phosphate coating is usually evaluated by analyzing the nature, shape and size of the crystals, and the distribution of crystals, evenness, and also by surface defects (pores, cracks) [3]. This technique reveals the characteristic properties of the crystal structure, the size of the crystallites, the extent and evenness of the coating, the parameters that determine coating performance.

The structure of the phosphate coats was analyzed using a Zeiss Imager Axio a1M microscope and a Vega Tescan LMH II scanning electron microscope.

Fig. 4.5 shows the images recorded by light microscopy, at a magnification of X20 and X40, and Fig. 4.6 shows the SEM micrographs for the zinc phosphate and manganese coats deposited on carbon steel.

Iron oxides are detected on the C45 carbon steel, resulting from the contact between the iron ion and the humidity in the atmosphere, causing the dissociation of water molecules with the formation of iron oxide and release of hydrogen.

The zinc phosphate coat on the $I - Zn$ și $II - Zn/Fe$ samples occurs by the formation of trizinc phosphate tetrahydrate crystals. It consists of numerous crystals of different sizes that intertwine on the entire steel surface. The layer also has certain channels called intercrystalline areas with specific porosity characteristics [4,5]. The phosphate crystals present on the steel surface are fine, compact and flower or plate-shaped, this shape being specific to hopeite. The surface structure of the base metal determines the orientation of the phosphate crystals and hence the structure of the deposited phosphate coat [6]. The main difference between the two types of samples is the size of the phosphate crystal, which is much smaller in the second sample.

The deposited coats have a rough, porous and plate-like surface structure. These pores have an irregular geometry structure and can penetrate the entire cross section of the phosphate coat. Another interesting observation is the presence of bright lines in the coatings in images with SEM section. EDX analysis reveals a higher ratio of zinc to phosphorus (and oxygen) in these locations [7].

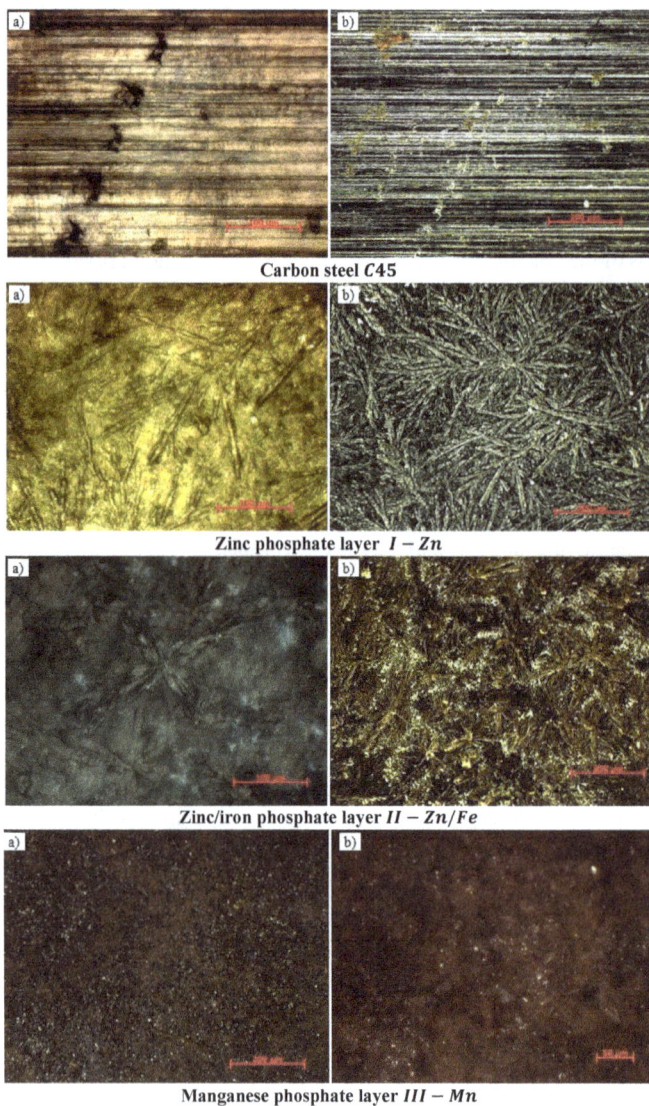

Figure 4.5 *Optical microstructure of carbon steel and phosphate layers deposited on its surface at magnification: a) 40x; b) 20x.*

Zinc phosphate layer $I - Zn$

Zinc/iron phosphate layer $II - Zn/Fe$

Manganese phosphate layer $III - Mn$

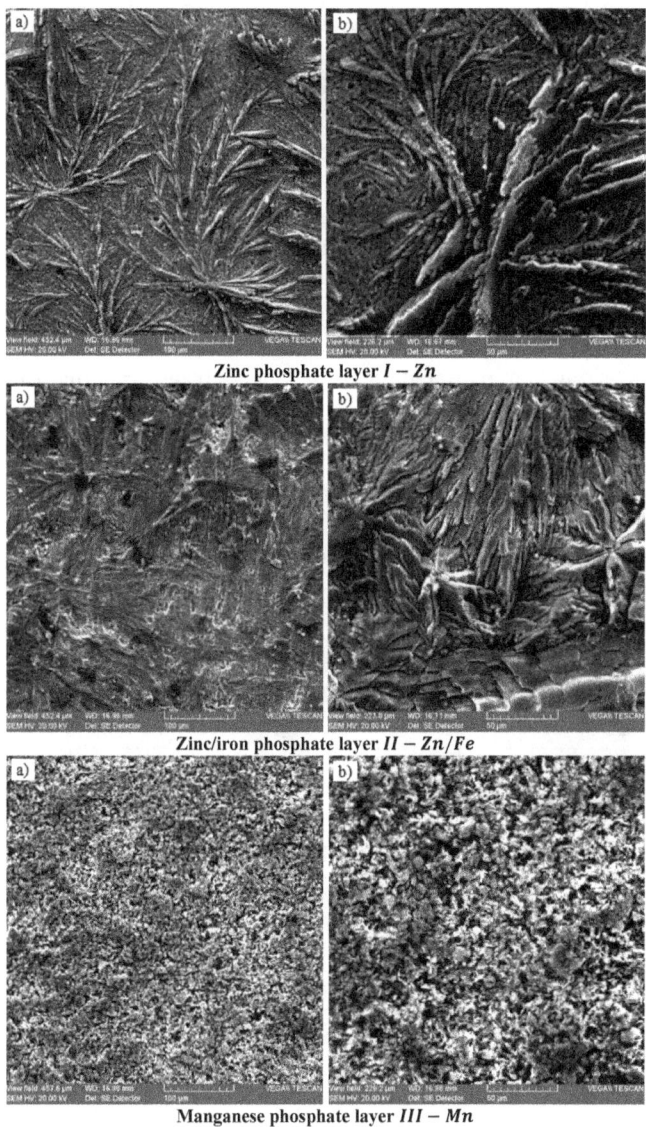

Figure 4.6 Morphology of phosphate layers at different magnifying powers: a)100X, b) 1000X.

The manganese phosphate layer has a different morphology compared to the zinc phosphate and zinc/iron phosphate coats. The crystals corresponding to zinc phosphate are larger compared to those of manganese phosphate. The small size of the crystals formed on the steel surface and the porous crystalline structure are specific to manganese phosphate. By adding nickel and iron to the manganese phosphate solution, the size of the phosphate crystals decreases, with large areas evenly covered as a result of crystal overlapping at microscopic level. This change in the phosphate layer may be attributed to the accelerating effect of galvanic torques formed between Fe and Ni ions, which act as crystallization centers in the formation process, increasing their number exponentially [8].

4.3 Characterization of phosphate coats by Fourier transform infrared spectroscopy

Structural characterization by Fourier transform infrared spectroscopy (FTIR) was performed to detect the main compounds specific to phosphate coats [9,10]. At the same time, this analysis can confirm the formation of the coating and its degree of hydration.

Fourier transform infrared spectroscopy tests were done using a Bruker Hyperion 1000 FTIR spectrometer, in a range of wave numbers between 4000 cm^{-1} and 600 cm^{-1} at a spectral resolution of 4 cm^{-1} and a number of 64 scans for each analyzed surface. The FTIR spectrum of the zinc phosphate coated sample, $I - Zn$, represented graphically by the wave number on the abscissa and the intensity of the infrared radiation in absorbance on the ordinate is shown in Fig. 4.7.

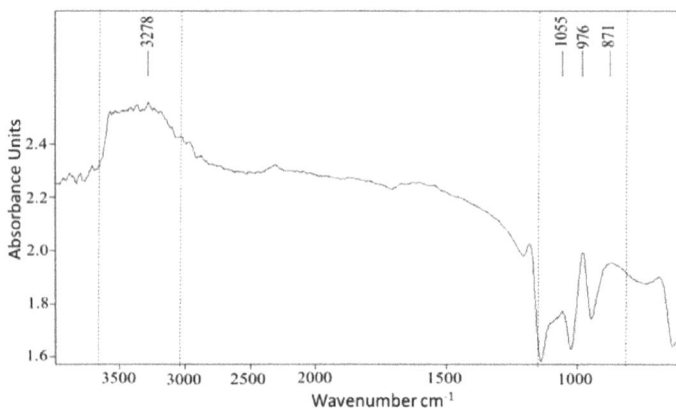

Figure 4.7 FTIR spectrum of zinc phosphate layer.

As one may notice in Fig. 4.7, the vibration band ranging between 1100 and 850 cm-1 has several inflection points of significant intensity. Their positioning at 871, 976 and 1055 cm^{-1} may be attributed to the asymmetric valence vibration of the reaction products between phosphate and hydroxide in the form of $P - O$ or $P - O - H$ groups from the complex chemical compounds $HOPO_3^{3-}$, PO_4^{3-} și $H_2PO_4^-$ specific to phosphate coats deposited by conversion [11].

The FTIR spectrum of the iron and zinc phosphate coated sample, $II - Zn/Fe$, shows three vibration bands as in Fig. 4.8.

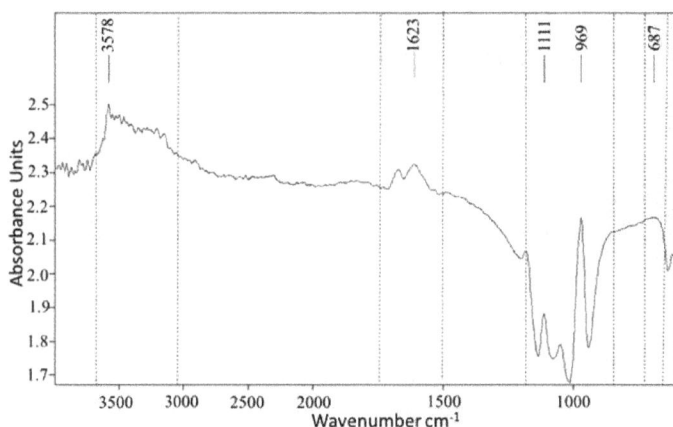

Figure 4.8 FTIR spectrum of zinc-iron phosphate layer.

Due to the similar chemical composition of the two zinc-based phosphating solutions, the FTIR spectra are somewhat similar. The vibration band ranging between 710 cm^{-1} and 680 cm^{-1} shows an inflection point at the wave number 687 cm^{-1} that may be attributed to the asymmetric valence vibrations of the metaphosphate groups. The occurrence of peaks in the vibration band ranging between 1700 and 1500 cm^{-1} corresponds to the adsorbed water molecules, which highlights the degree of hydration of the zinc-iron phosphate coat [11]. A vibration band ranging between 3600 and 3100 cm^{-1} corresponding to the valence vibrations of the hydroxyl groups, and a band ranging from 1100 to 850 cm^{-1}, which is attributed to the vibrations of phosphate ions, are also present [12].

The large width of the vibration band ranging between 1100 and 850 cm^{-1} is due to the different groups of phosphates in the structure, the large number of vibrations in this area being due to the connections between these groups [13].The 1111 cm^{-1} peak may be attributed to the asymmetric vibrations of PO_4^{3-} valence in in the phosphophyllite structure, whereas the 969 cm^{-1} peak may be caused by the deformation vibrations of the $P - O$ groups [14].

The FTIR spectrum shows major changes due to the depositing of a manganese coat on the samples, $III - Mn$ (Fig. 4.9).

The inflection point at 918 cm^{-1} on the vibration band ranging between 1100 and 850 cm^{-1} confirms the formation of manganese phosphate by the presence of $P - O$ and $P - O - H$ groups in $Mn_5Fe_2(PO_4)_4.4H_2O$ [1]. The 3386 cm^{-1} peak is attributed to the hydroxyl groups in complex chemical compounds [15]. The band ranging between 1700 and 1500 cm^{-1} is attributed to the deformation vibrations of the $H - O - H$ groups in the adsorbed water molecules [16].

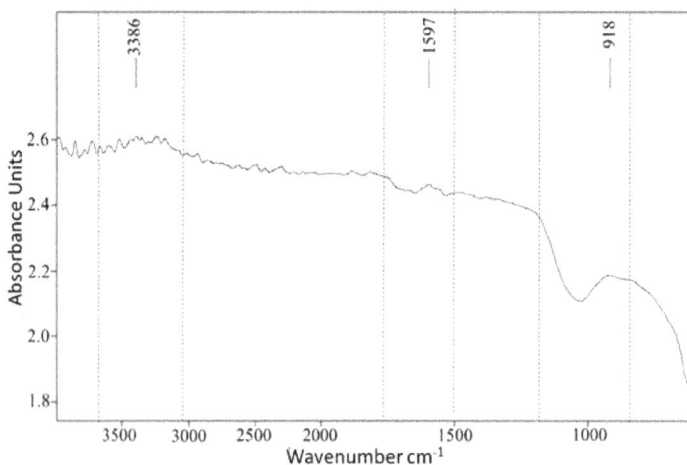

Figure 4.9 FTIR spectrum of manganese phosphate layer.

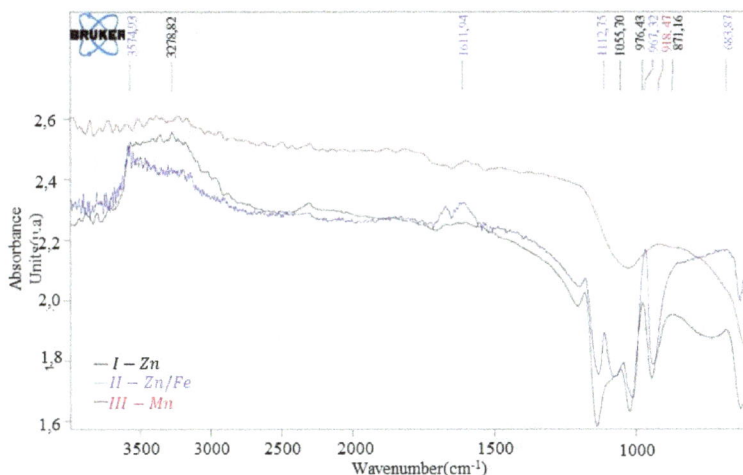

Figure 4.10 FTIR spectrum of phosphate layers.

In the case of both zinc and manganese phosphate coats, the spectrum has a vibration band ranging between 1100 and 850 cm^{-1}, which is attributed to the $P - O$ and $P - O - H$ groups. On this vibration band, the $I - Zn$ sample shows a high intensity peak positioned at 976 cm^{-1} which confirms the formation of hopeite, but after the introduction of iron in the phosphating solution the main compound formed is phosphophyllite. On the other hand, by replacing zinc with manganese, the number of $P - O$ vibrations decrease significantly, resulting in a more porous layer with reduced crystallinity (as shown in Fig. 4.10).

The FTIR spectra of the phosphate coats deposited by conversion show some bands of vibrations similar to those of hydroxide groups, and also bands positioned differently that reveal the specificity of the compounds formed [17]. In the case of both zinc phosphate and manganese coats, the spectrum has a vibration band ranging between 1100 and 850 cm^{-1} which is attributed to the $P - O$ and $P - O - H$ groups.

References

[1]   A.V. Sandu, C. Coddet, C. Bejinariu, A Comparative Study on Surface Structure of Thin Zinc Phosphates Layers Obtained Using Different Deposition Procedures on Steel, REVISTA DE CHIMIE, 63(4) (2012) 401-406.

[2]   A.V. Sandu, A. Ciomaga, G. Nemtoi, C. Bejinariu, I. Sandu, SEM-EDX and

microFTIR studies on evaluation of protection capacity of some thin phosphate layers, Microscopy Research and Technique, 75(12) (2012) 1711-1716. https://doi.org/10.1002/jemt.22120

[3] C. Bejinariu, P. Lazăr, A.V. Sandu, A.M. Cazac, I.G. Sandu, O. Corbu, Enhancing properties of reinforcing steel by chemical phosphatation, APPLIED MECHANICS AND MATERIALS, 754-755 (2015) 310-314. https://doi.org/10.4028/www.scientific.net/AMM.754-755.310

[4] C. Bejinariu, D.P. Burduhos-Nergis, N. Cimpoesu, M.A. Bernevig-Sava, S.L. Toma, D.C. Darabont, C. Baciu, Study on the anticorrosive phosphated steel carabiners used at personal protective equipment, Quality-Access to Success 20(1) (2019) 71-76.

[5] D.P. Burduhos-Nergis, C. Nejneru, R. Cimpoesu, A.M. Cazac, C. Baciu, D.C. Darabont, C. Bejinariu, Analysis of Chemically Deposited Phosphate Layer on the Carabiners Steel Surface Used at Personal Protective Equipments, Quality-Access to Success 20(1) (2019) 77-82.

[6] A.V. Sandu, A. Ciomaga, G. Nemtoi, C. Bejinariu, I. Sandu, Study on the chemical deposition on steeel of zinc phosphate with other metallic cations and hexamethilen tetramine. II. Evaluation of corrosion resistance, Journal of Optoelectronics and Advanced Materials, 14(7-8) (2012) 704-708.

[7] D.P. Burduhos-Nergis, P. Vizureanu, A.V. Sandu, C. Bejinariu, Evaluation of the Corrosion Resistance of Phosphate Coatings Deposited on the Surface of the Carbon Steel Used for Carabiners Manufacturing, Applied Sciences 10(8) (2020) 2753. https://doi.org/10.3390/app10082753

[8] J. Duszczyk, K. Siuzdak, T. Klimczuk, J. Strychalska-Nowak,A. Zaleska-Medynska, Manganese Phosphatizing Coatings: The Effects of Preparation Conditions on Surface Properties, Materials 11(12) (2018) 2585. https://doi.org/10.3390/ma11122585

[9] T.S.N. Sankara Narayanan, Surface pretreatment by phosphate conversion coatings - a review. Rev.Adv.Mater.Sci., 9 (2005) 130-177.

[10] A.V. Sandu, C. Bejinariu, A. Predescu, I.G. Sandu, C. Baciu, I. Sandu, New mechanisms for chemical phosphatation of iron objects, RECENT PATENT ON CORROSION SCIENCE, (ISSN 1877-6108), Bentham Science Publishers, 1(1) (2011) 33-37. https://doi.org/10.2174/2210683911101010033

[11] N.A. Ghoneim, A.M. Abdelghany, S.M. Abo-Naf, F.A. Moustafa, Kh.M. ElBadry,

Spectroscopic studies of lithium phosphate, lead phosphate and zinc phosphate glasses containing TiO2: Effect of gamma irradiation. J.Mol.Struct., 1035 (2013) 209-217. https://doi.org/10.1016/j.molstruc.2012.11.034

[12]   Y. Otsuka, M. Takeuchi, M. Otsuka, B. Ben-Nissan, D. Grossin, H. TanakaEffect of carbon dioxide on self-setting apatite cement formation from tetracalcium phosphate and dicalcium phosphate dihydrate;ATR-IR and chemoinformatics analysis. Colloid Polyn Sci., 293 (2015) 2781-2788. https://doi.org/10.1007/s00396-015-3616-6

[13]   Y.M. Moustafa, K. El-Egili, Infrared spectra of sodium phosphate glasses. JNCS, 240 (1998) 144 153. https://doi.org/10.1016/S0022-3093(98)00711-X

[14]   J.D. Wang, D. Li, J.K. Liu, X.H. Yang, J.L. He, Y. Lu, One-Step Preparation and Characterization of Zinc Phosphate Nanocrystals with Modified Surface. SNL, 1 (2011) 81-85. https://doi.org/10.4236/snl.2011.13015

[15]   E. Alibakhshi, E. Ghasemi, M. Mahdavian, B. Ramezanzadeh, Corrosion Inhibitor Release from Zn-Al-[PO43-]-[CO32-] Layered Double Hydroxide Nanoparticles. Prog. Color Colorants Coat., 9 (2016) 233-248.

[16]   D.P. Burduhos-Nergis, A.M. Cazac, A. Corabieru, E. Matcovschi, C. Bejinariu, Characterization of Zinc and Manganese Phosphate Layers Deposited on the Carbon Steel Surface, Euroinvent ICIR IOP Conference Series: Materials Science and Engineering 877 (2020) 012012. https://doi.org/10.1088/1757-899X/877/1/012012

CHAPTER 5

Mechanical Characterization of Coats

As zinc phosphate coats provide better durability of painted parts exposed to corrosive environments, the zinc phosphate achieved from the first phosphating solution was chosen as the base coat for the mechanical characterization of painted phosphate-coated steel.

5.1 Determination of mechanical properties by scratch and microindentation tests

The scratch and microindentation tests were performed using the equipment called Universal Micro-Tribometer.

The scratch test consists of slowing moving a micro-blade through the coating or the surface of the material [1]. The parameters of the scratch test are: the distance traveled by the blade, equal to 10 mm; time required to cover that distance, 60 seconds; blade speed of 167 μm/second. The material is gradually removed from the sample surface. This test measures the following parameters, Fx corresponding to the value of the friction force, Fz corresponding to the normal force and COF corresponding to the friction [2,3]. In this test, the normal force increases with increasing time from 0 to 10 N.

The microindentation test consists of two main steps: an initial preload step with a force of 1% of the maximum test force and a loading step with a force of up to 450 N, depending on the set value and the material on which the test is performed. The preload step aims to establish the reference point 0 of the graph, which is followed by an increase in force, constant over time, up to the maximum set value. Once this value has been reached, the load force is kept constant for a short time. This is followed by a decrease in the load force, known as discharge, which is achieved at constant rate over a predefined time [2,4].

The hardness measured by the indentation process is defined as the average contact pressure, H_{IT} and is calculated according to the following ratio:

$$H_{IT=}F_{max} \ / \ AP \qquad\qquad (5.1)$$

where Fmax is the maximum force, and AP is the projection of the contact area at this force.

The curves specific to indentation tests are achieved between load force variation, measured in N, and material deformation, measured in μm, as well as discharge force

Phosphate Coatings Suitable for Personal Protective Equipment   Materials Research Forum LLC
Materials Research Foundations **89** (2021)   https://doi.org/10.21741/9781644901113

variation. Irregularities shown by the loading curve are associated with the presence of cracks in the tested material to which the force is applied [5]. In some cases, irregularities may also occur on the discharge curve, which are caused by the material adhering to the penetrator. The graphical distance on the abscissa between point 0 and the test end point represents the residual material deformation after the test [6].

5.1.1   Scratch tests

Initial sample made of C45 steel (C45)

The C45 sample graph (Fig. 5.1) shows small variations in the value of the force and friction coefficient, graphically represented by peaks, up to about 55 seconds corresponding to a load of about 9.2 N, after which a sudden increase of specific curve slopes occurs that may be linked to the material getting stuck in front of the blade and resistance to its removal following a possible hardening.

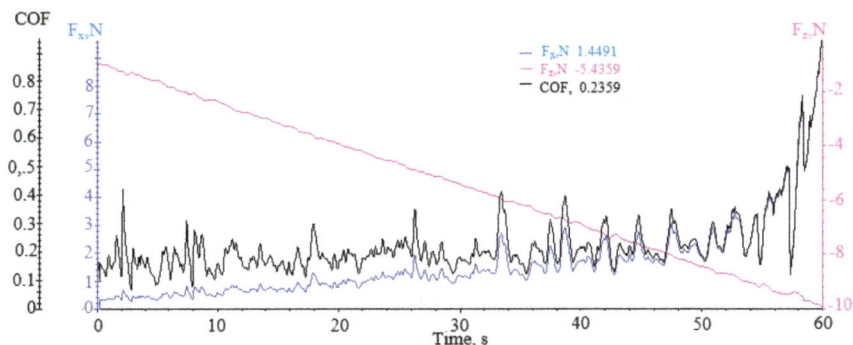

Figure 5.1 The C45 initial sample scratch test plot.

The mean value of the friction coefficient in this sample is 0.2359, while the standard deviation is 0.1152. The mean value of the friction force was 1.449 N with a standard deviation of 1.387.

The microstructural analysis of the samples (Fig. 5.2 a) highlights the specific scratch mark profile after the scratch test (Fig. 5.2 c). The 3D profile reveals, on the edge of the groove cut by the blade, two protrusions formed by the removed material (Fig. 5.2 d). In some areas there are also pieces of material stuck to the bottom of the groove (Fig. 5.2 b). The scratch mark is deeper than 100 μm.

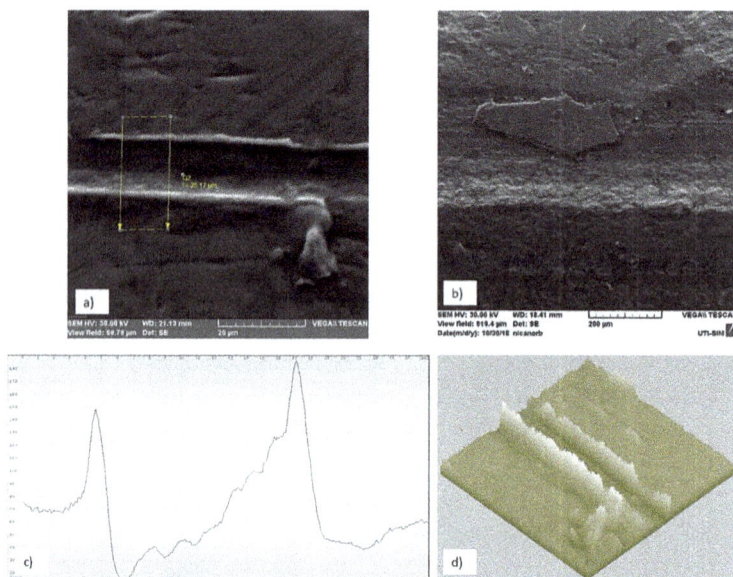

Figure 5.2 C45 sample microstructure: a) SEM micrograph of scratch track profile 3kx; b) SEM micrograph of stuck material zone 250x; c) scratch profile; d) 3D sample profile.

Phosphate-coated sample made of C45 steel

Due to the phosphate-coating of C45 steel samples, the mean value of the friction coefficient increased to 0.5915, while the standard deviation was 0.2361, and the friction force increased to a mean value of 3.738 N with a standard deviation of 2.568. This increase is due to the improvement of the mechanical properties following the deposition by conversion of the insoluble phosphate coat. The slopes specific to the friction coefficient and friction force curves (Fig. 5.3) show a sudden change after about 39 seconds, which corresponds to a load value of about 6.5 N. Thus, it can be seen that following the phosphating process the sample shows better scratch resistance and a phosphate coat with a friction coefficient higher than that of the base material.

The microstructural analysis of the phosphate coat deposited by conversion on the surface of C45 steel samples reveals the mark of the blade after the scratch test (Fig. 5.4 a) and its profile (Fig. 5.4 b).

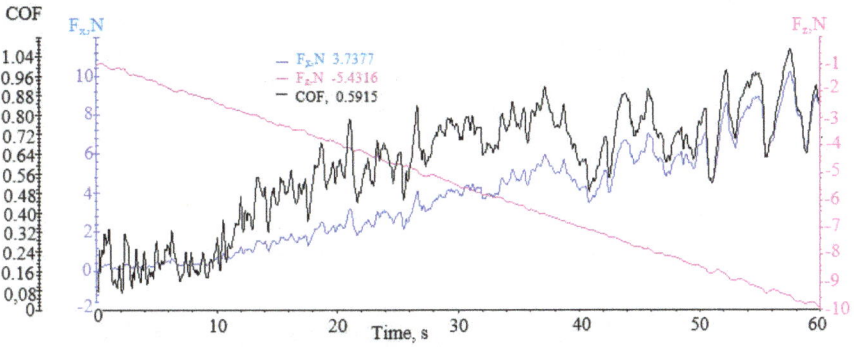

Figure 5.3 The I − Zn sample scratch test plot.

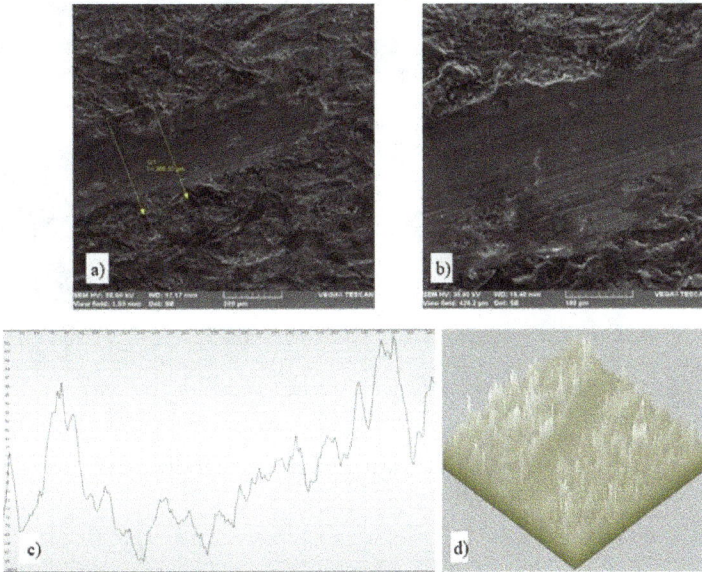

Figure 5.2 I − Zn sample microstructure: a) SEM micrograph of scratch track profile 3kx; b) SEM micrograph of stuck material zone 500x; c) scratch profile; d) 3D sample profile.

As shown by the 3D analysis (Fig. 5.4 c), the sample has a rough surface with many protrusions specific to crystal microdendrites. As in the case of non-phosphate-coated samples, the material has been either torn or stuck to the bottom of the groove in some areas, caused by the passage of the blade (Fig. 5.4 d).

Elemental surface mapping using EDX detects the areas where the coating was pierced by the blade, reaching the base material (Fig. 5.5). The percentage of Fe is directly proportional to the exfoliated surface of the coating. In areas where Fe occupies the entire width of the blade mark, the phosphate coat has been completely removed.

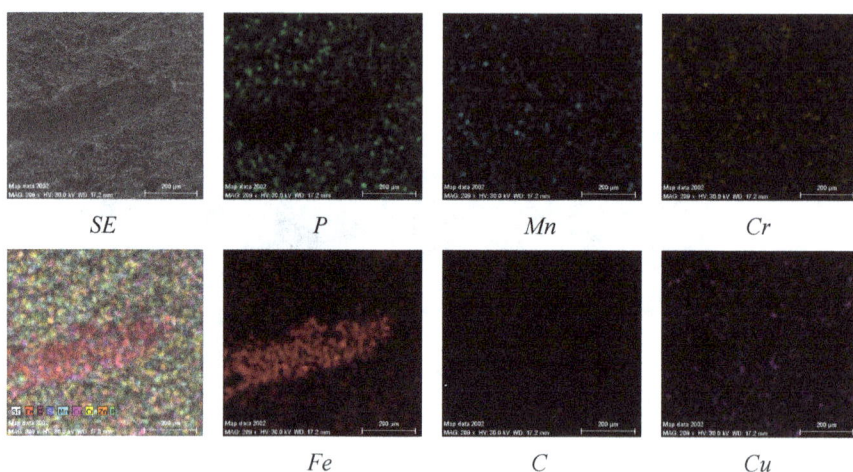

SE	*P*	*Mn*	*Cr*
	Fe	*C*	*Cu*

Figure 5.5 Main elements mapping of I − Zn phosphate samples after scratch test.

The phosphate coat deposited on the steel surface is extremely scratch resistant, as it only removed in the final part of the test.

Phosphate-coated sample made of C45 steel impregnated in lubricant (OFU)

In order to reduce the apparent friction coefficient, the outer porous layer may be impregnated with different types of lubricants. The effect of their impregnation in a lubricant containing MoS_2 and having superior lubrication characteristics was studied. The mean friction coefficient of the phosphate-coated C45 sample impregnated in lubricant was 0.2955, while its standard deviation was 0.0755 (as one may note, the fluctuation range of the peaks was much narrower, a phenomenon that could be accounted for by the easier and more smooth sliding of the blade during testing, in other words due to the lubricant). The graph specific to this type of sample (Fig. 5.6) shows fluctuations linked to the phosphate structure. However, there is no significant change in the curve slope, as the blade did not seem to have penetrated the phosphate coat.

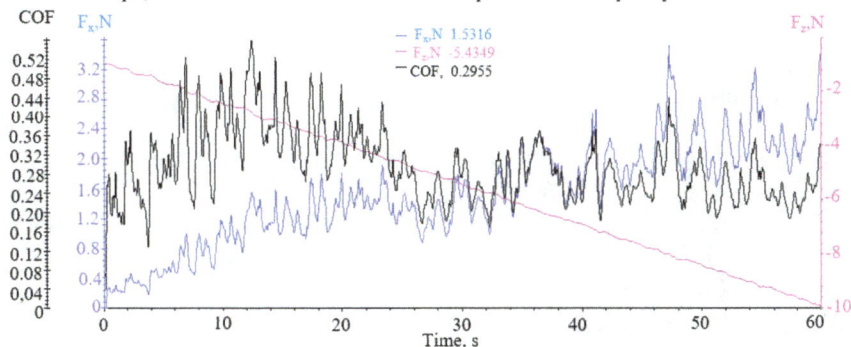

Figure 5.6 The OFU sample scratch test plot.

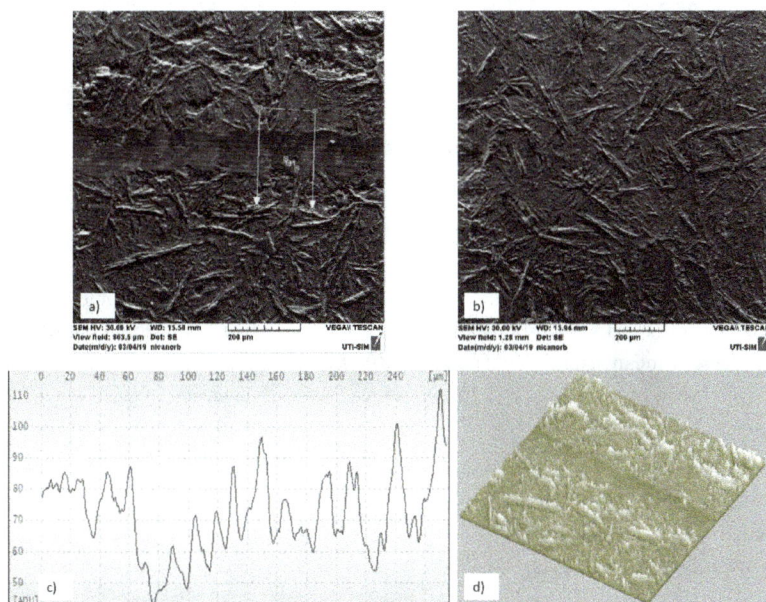

Figure 5.7 OFU sample microstructure: a) SEM micrograph of scratch track profile 3kx; b) SEM micrograph of stuck material zone 500x; c) scratch profile; d) 3D sample profile.

The microstructural analysis of the surface of the sample (Fig. 5.7 b) impregnated in lubricant reveals the mark of the blade after the scratch test (Fig. 5.7 a) and its profile (Fig. 5.7 c). The 3D profile of the sample has a relatively flat surface due to the pores filled with lubricant that flatten the 3D surface (Fig. 5.7 d).

As shown by elemental surface mapping (Fig. 5.8), the blade did not penetrate the phosphate coat during the scratch test carried out according to the same parameters on the phosphate-coated sample impregnated in lubricant.

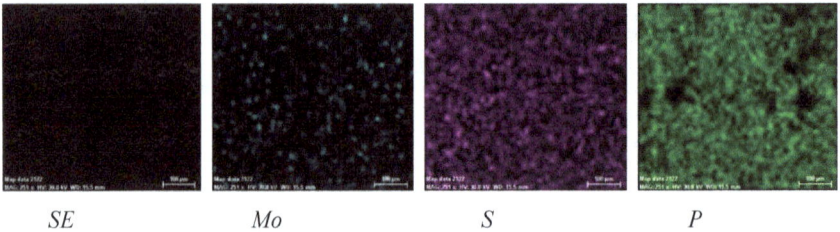

SE   Mo   S   P

Figure 5.8 Main elements mapping of OFU sample after scratch test.

Elemental mapping shows an even distribution of the chemicals included in the molybdenum disulphide-based lubricant used for impregnation. Areas rich in molybdenum (Mo) and sulfur (S) are specific to large pores.

Painted sample made of C45 steel (C45V)

Since cracks may occur in the material that carabiners are made of following their hitting rocks or falling from various heights, which may have dramatic consequences during their use, coating them with elastomer-based paint may be a smart solution to protect carabiners from shocks. Non-phosphate-coated samples on which a thin coat of paint is applied have a mean friction coefficient of 0.6922 and a standard deviation of 0.2942, while their mean friction force is 3.156 N, with a standard deviation of 0.9076.

Fig. 5.9 shows that after about 13 seconds, i.e. when the load reaches 3 N, the blade penetrated the elastomer-based paint coat, which decreased the friction coefficient curve. The friction force curve has an upward trend throughout the test, which may be accounted for by the adhesion of the elastomer-based paint to the surface of the blade, thus preventing it from moving forward.

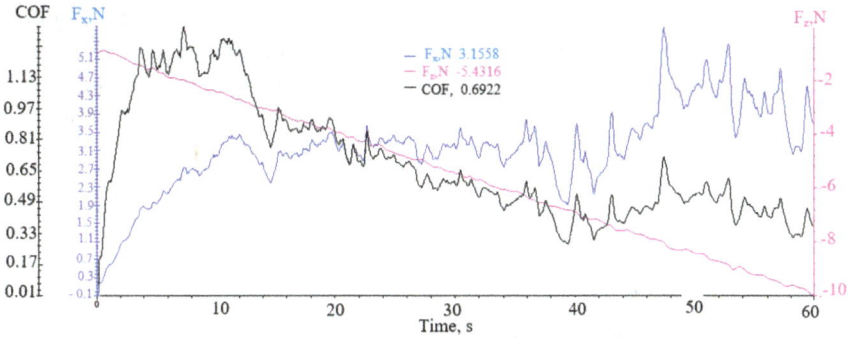

Figure 5.9 The C45V sample scratch test plot.

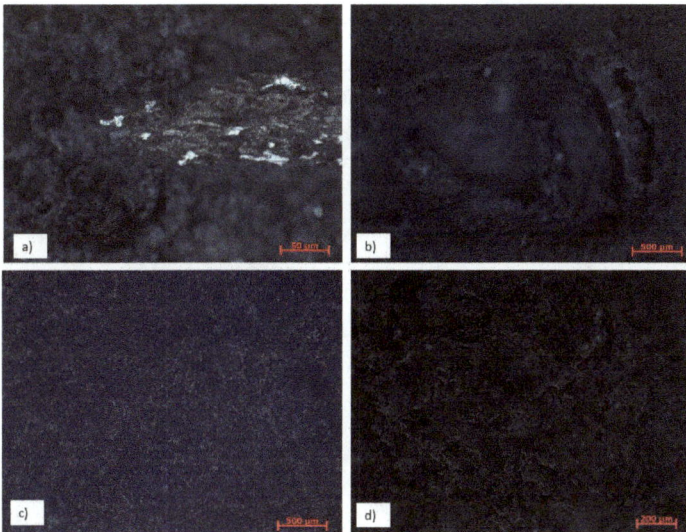

Figure 5.10 C45V microstructure: a) optical micrograph of scratch track profile 100x (track start); b) scratch track optical micrograph 100x (trak end); c) rubber paint layer optical micrograph 100x (bright field); d) rubber paint layer optical micrograph 100x (dark field).

The mark left by the blade during the scratch test may be analyzed using light microscopy. Complete mark analysis may reveal that at its beginning (Fig. 5.10 a) a large part of the elastomer paint coat is removed, yet there are still significant areas covered by the paint due to its specific adhesion properties. As the blade moves forward, the influence of the paint coat decreases considerably. Towards the end (Fig. 5.10 b) of the mark, one may note that the elastomer coat did not break due to its elastic properties, stretching significantly before breaking. The paint coat sprayed on the sample is evenly distributed on the carbon steel surface (Fig. 5.10 c, d).

Painted phosphate-coated sample made of C45 steel (OFV)

Following the phosphate-coating process, a coat of phosphates with high porosity is deposited on the surface of the sample, which helps the elastomer-based paint to better adhere to the surface. The C45 steel sample coated with phosphates and paint has a mean friction coefficient of 1.715 with a deviation of 0.5392, and a mean friction force of 8.562 N with a deviation of 3.817.

The friction coefficient curve shows a sudden decrease about 10 seconds later, i.e. when the load is about 3N, which may be correlated with the penetration of the paint coat by the blade followed by an increase due to the penetrator touching the phosphate coat (Fig. 5.11). The blade penetrates the phosphate coat about 44 seconds later, when the load reaches about 7N.

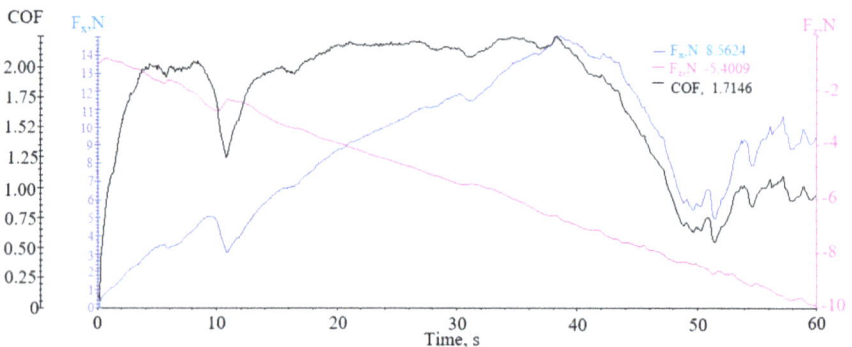

Figure 5.11 The OFV sample scratch test plot.

The microstructural analysis of the paint coat (Fig. 5.12 b) deposited on the phosphate surface reveals the mark left by the blade during the scratch test (Fig 5.12 a). The scratch profile (Fig. 5.12 c) shows two high peaks, which are closely dependent on the amount of paint removed, and several peaks between the two profiles, which represent the phosphate dendrites on the surface, which are also confirmed by the 3D profile of the sample (Fig. 5.12 d).

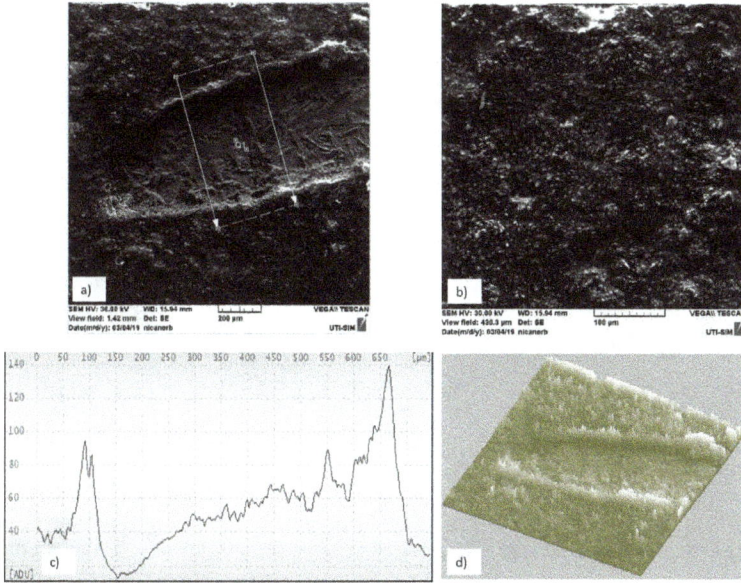

Figure 5.12 OFV sample microstructure: a) SEM micrograph of scratch track 200x; b) rubber paint layer SEM micrograph 100x; c) scratch profile; d) 3D sample profile.

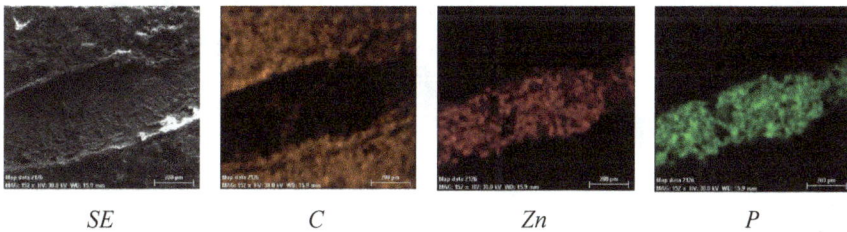

SE   C   Zn   P

Figure 5.13 Main elements mapping of OFV sample.

The elemental mapping of the scratch mark on the phosphate- and paint-coated sample shows the surfaces from which the protective coat against mechanical shocks has been partially removed. The presence of carbon (C) in the elemental mapping confirms the adhesion of the elastomer coat to the phosphate surface (Fig. 5.13).

Sample C45/ Sample I − Zn

The deposition of the phosphate coat on the surface of the C45 steel sample triggers changes in the friction coefficient curve determined during the scratch test (Fig. 5.14). The friction coefficient value increases due to sample phosphate-coating from about 0.2359 to about 0.5915, i.e. by about 150%.

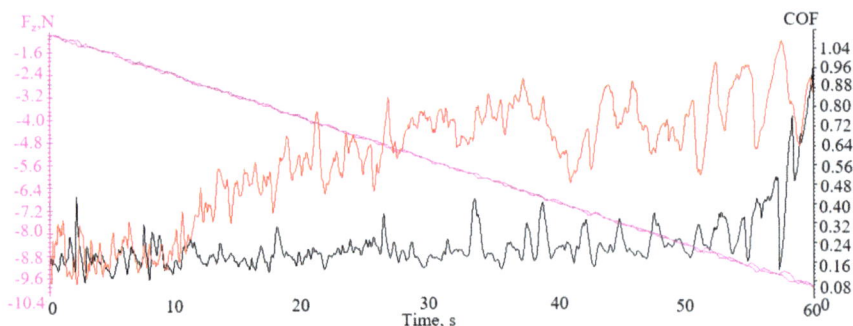

Figure 5.14 The C45 and I − Zn samples scratch test plot.

Sample I − Zn/ Sample OFU

By soaking the porous phosphate coat in MoS2-based lubricant, the mean friction coefficient of the phosphate-coated sample dropped by about 50%, namely from 0.5915 to less than 0.2955. The effect of the lubricant on the COF is significant, as it reduces the mean value of the phosphate-coated sample to a value similar to a non-phosphate-coated sample (Fig. 5.15).

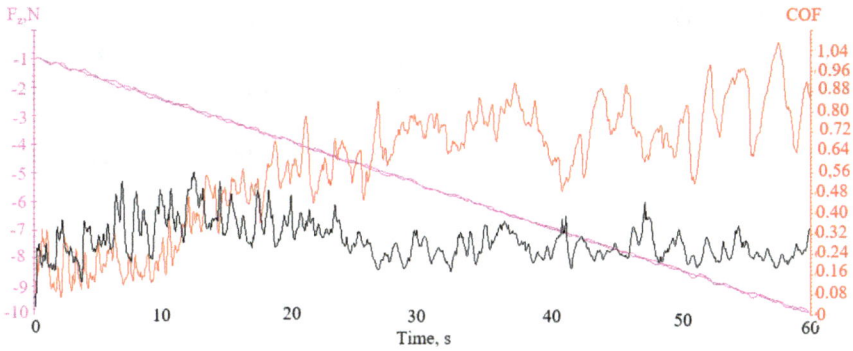

Figure 5.15 The I − Zn and OFU samples scratch test plot.

Sample C45V / Sample OFV

After a coat of paint has been applied to phosphate-coated samples, we noted an approximately 50% increase of the friction coefficient, i.e. from about 0.6922, a value specific to the painted non-phosphate-coated sample, to about 1.715, which corresponds to the painted phosphate-coated sample (Fig. 5.16).

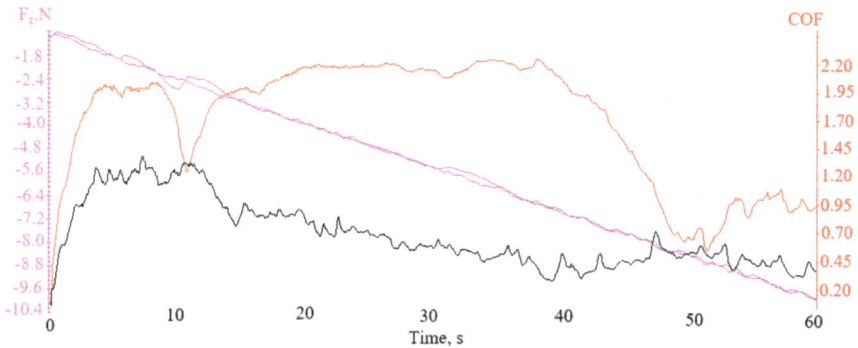

Figure 5.16 The C45V and OFV samples scratch test plot.

5.1.2 Microindentation test

The microindentation test was performed using the CETR UMT-2 tribometer and the Rockwell method. The initial preload stage, carried out with a force of 1% of the maximum test force, is achieved within 10 s to 20 s. The maximum load force used in this study was 15 N. It is achieved by even increase after setting the moment 0 for 30 seconds. After this value has been reached, the load force is maintained for 20 s, followed by the discharge period (Fig. 5.17).

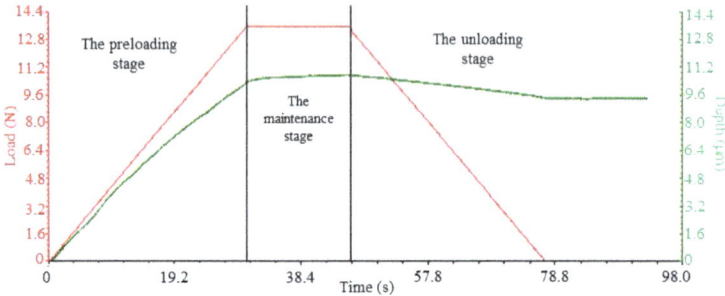

Figure 5.17 The microindentation stages.

Initial sample made of C45 steel (C45)

Three tests were performed on the C45 non-phosphate-coated steel sample, the results being shown in Table 5.1. The mean hardness is 15.35 GPa, the modulus of elasticity is 367.02 GPa, while the mean maximum penetration depth is 9.69 μm.

Table 5.1 Microindentation test results for the C45 sample.

Indentation number	Hardness, [GPa]	Elastic modulus, [GPa]	Depth, [μm]
1	17.55	386.36	9.04
2	14.19	364.65	10.03
3	14.10	350.07	10.06
Average	15.28	367.03	9.71

100

The graph of the sample with the values closest to the average ones (Fig. 5.18) shows a residual deformation of about 9.4 μm, as the system's behavior is plastic and no microcracks occur. The value of the elasticity of the material is determined by subtracting the maximum penetration depth from the residual deformation, which in the initial sample is 0.63 μm (about 6%).

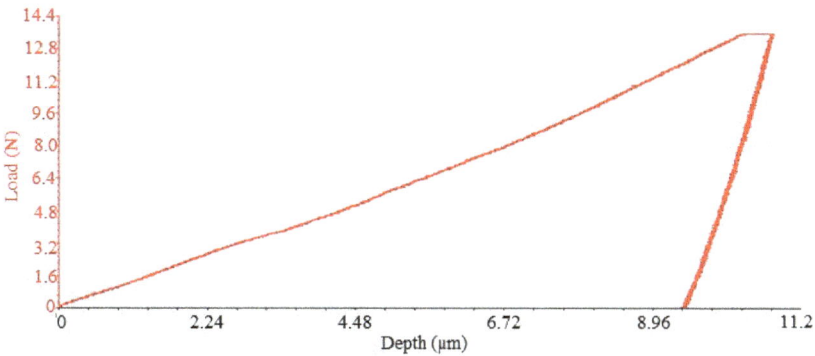

Figure 5.18 The C45 sample microindentation test chart.

Phosphate-coated sample made of C45 steel (I-Zn)

The phosphate coat deposited by conversion on the surface of the C45 steel sample has microhardness characteristics similar to the original sample. The mean coat hardness is 13.12 GPa, i.e. by about 15% lower, with a specific modulus of elasticity of 388.95 GPa, higher than that of the initial sample (Table 5.2). The significant differences between these values may be due to the inhomogeneity of the thickness of the deposited coat.

Table 5.2 Microindentation test results for the I − Zn sample.

Indentation number	Hardness, [GPa]	Elastic modulus, [GPa]	Depth, [μm]
1	15.16	402.65	9.73
2	13.30	401.68	10.31
3	10.91	362.54	11.45
Average	13.12	388.95	10.49

The analysis of the phosphate-coated sample graph (Fig. 5.19), which is closest to the mean values, reveals that the phosphate coat has an elastoplastic behavior with a residual deformation of about 9.8 µm. During discharge, that variation (solid line) is different from the elastic back part (dotted line of force), which suggests the presence of plastic deformation. During the loading stage, the graph shows small fluctuations due to the microcracking processes that take place under external load.

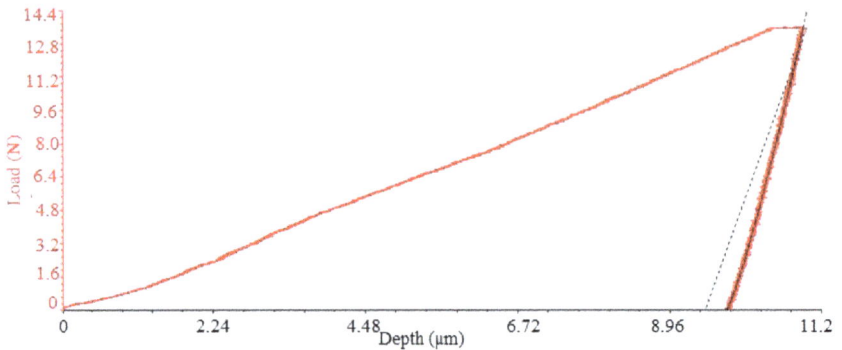

Figure 5.19 The I − Zn sample microindentation test chart.

Initial painted sample made of C45 steel (C45V)

The coat of paint sprayed on the surface of the samples has very low hardness values, which are specific to elastomeric materials. The mean maximum penetration depth for a maximum load peak of 15 N is about 47 µm (Table 5.3).

Table 5.3 Microindentation test results for the C45V sample.

Indentation number	Hardness, [GPa]	Elastic modulus, [GPa]	Depth, [µm]
1	0.45	125.69	56.07
2	0.82	156.77	41.58
3	0.67	138.48	44.71
Average	0.65	140.31	47.45

From a graphical point of view, the elastomer-coated sample shows a loading curve specific to elastic materials followed by a discharge curve that characterizes plastics (Fig. 5.20). This is due to the high loading force used during the test, which causes cracks in the paint coat at great depths that advance faster than the penetrator. The difference between the maximum penetration depth and the residual deformation (about 44 μm) is very small, i.e. about 1%.

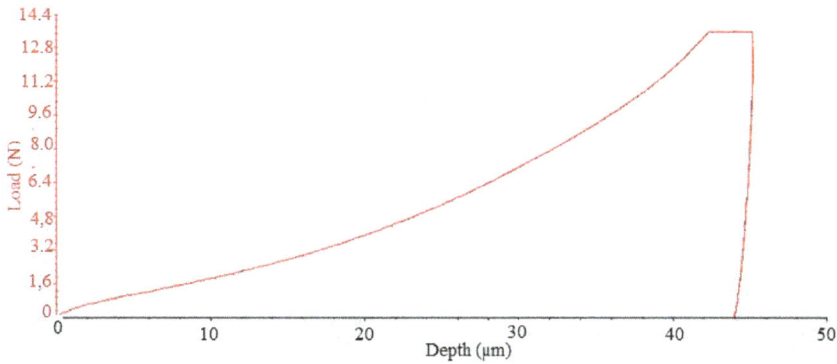

Figure 5.20 The C45V sample microindentation test chart.

5.2 Determination of impact resistance

The insoluble phosphate coat provides high protection against corrosion, while the paint coat can absorb mechanical shocks that may cause internal cracks [7-14]. Coats of black elastomer-based paint are a cheap, simple and fast method of protection against mechanical shocks for C45 steel carabiners. The outer layer has high elasticity, does not crack, does not crumble and adheres very well to the porous surface of the phosphate coat [15].

Impact resistance was determined by means of a pendulum for bending by shock - Charpy hammer, according to standard EN 10045-1: 1990, on specimens with the dimensions specified in Fig. 5.21.

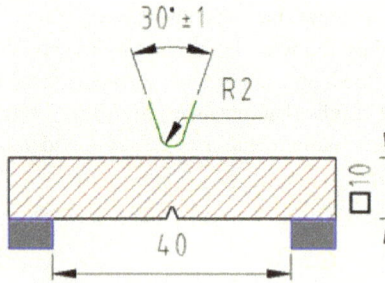

Figure 5.21 The dimensions of the tested samples.

As shown in Figs. 5.22 (b) and 5.23 (a, b), the phosphate coat has a porous surface, due to the formation of intertwined crystalline dentrites, which has the advantage of allowing coats of different types of paints [16-18].

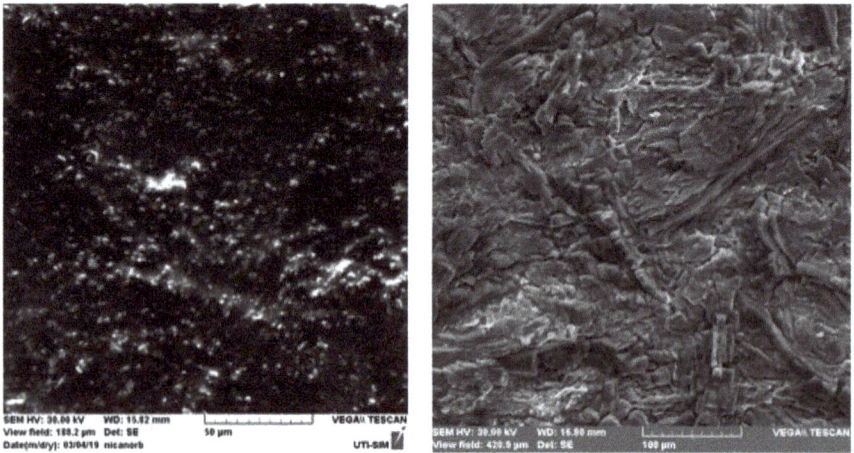

Figure 5.22 SEM micrographs of layers: a) Rubber paint layer SEM micrography, 50 μm; b) Phosphate layer SEM micrography, 100 μm.

The painted samples have an even, relatively porous surface as shown in Figs. 5.22 (a) and 45.23 (c, d), which may mean a high friction coefficient. It is therefore recommended that elastomer-based paint coating be done only on the outer surface of the carabiner.

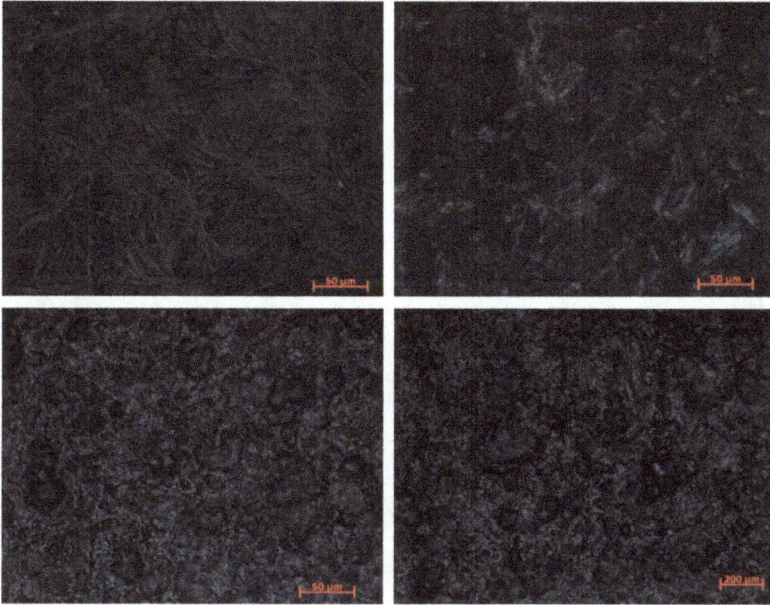

Figure 5.23 Optical microstructure: a, b) phosphate layer microstructure, X50; c) rubber painted layer microstructure, X50; d) rubber painted layer microstructure, X200.

Phosphate coating positively influences the shock resistance of the specimens, as it improves from 166 KJ, which is the value specific to the initial sample, to 182 KJ. Samples coated with an elastomer-based paint show the most significant increase, from 182 KJ to 209 KJ, as shown in Fig. 5.24.

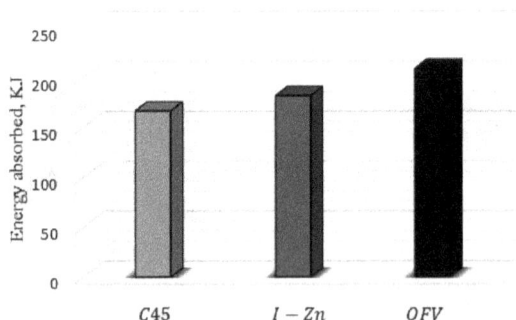

Figure 5.24 Tested samples toughness.

Phosphate-coated samples have an impact resistance about 9% higher than the initial samples, also phosphate-coated samples painted with elastomer paint have an impact resistance 15% higher than that of phosphate-coated samples. Therefore, the impact resistance of both paint- and phosphate-coated samples is 26% higher than that of C45 samples, which demonstrates that the deposited coats provide high protection against mechanical shocks to coated objects.

References

[1]   D. Ernens, M.B. de Rooij, H.R. Pasaribu, E.J. van Riet, W.M. van Haaften, D.J. Schipper, Mechanical characterization and single asperity scratch behaviour of dry zinc and manganese phosphate coatings. Tribology International, 118 (2018) 474-483. https://doi.org/10.1016/j.triboint.2017.04.034

[2]   CETR UMT Multi-Specimen Test System, Hardware Manual, 2007.

[3]   D.P. Burduhos-Nergiş; N. Cimpoesu, P. Vizureanu, C. Baciu, C. Bejinariu, Tribological characterization of phosphate conversion coating and rubber paint coating deposited on carbon steel carabiners surfaces. Materials Today: Proceedings, 19 (2019) 969–978. https://doi.org/10.1016/j.matpr.2019.08.009

[4]   K. Khlifi, H. Dhiflaoui, L. Zoghlami, A. Ben Cheikh Larbi, Study of mechanical behavior, deformation, and fracture of nano-multilayer coatings during microindentation and scratch test. J. Coat. Technol. Res., 12(3) (2015) 513-524. https://doi.org/10.1007/s11998-015-9662-7

[5]   Y. Huang, Z. Xue, H. Gao, W.D. Nix, Z.C. Xia, A Study of Microindentation Hardness Tests by Mechanism-based Strain Gradient Plasticity. Journal of

Materials Research, 15(8) (2000) 1786-1796.
https://doi.org/10.1557/JMR.2000.0258

[6]   E. Broitman, Indentation Hardness Measurements at Macro-, Micro-, and
       Nanoscale: A Critical Overview. Tribology Letters, 65 (2016) 1-18.
       https://doi.org/10.1007/s11249-016-0805-5

[7]   D.P. Burduhos-Nergis, P. Vizureanu, A.V. Sandu, C. Bejinariu, Evaluation of the
       Corrosion Resistance of Phosphate Coatings Deposited on the Surface of the
       Carbon Steel Used for Carabiners Manufacturing, Applied Sciences 10(8) (2020)
       2753. https://doi.org/10.3390/app10082753

[8]   D.P. Burduhos-Nergis, P. Vizureanu, A.V. Sandu, C. Bejinariu, Phosphate Surface
       Treatment for Improving the Corrosion Resistance of the C45 Carbon Steel Used
       in Carabiners Manufacturing, Materials 13(15) (2020) 3410.
       https://doi.org/10.3390/ma13153410

[9]   D.P. Burduhos-Nergis, C. Nejneru, D.D Burduhos-Nergis, C. Savin, A.V Sandu,
       S.L Toma, C. Bejinariu, The Galvanic Corrosion Behavior of Phosphated Carbon
       Steel Used at Carabiners Manufacturing, Revista de chimie 70(1) (2019) 215-219.
       https://doi.org/10.37358/RC.19.1.6885

[10]   C. Bejinariu, D.P. Burduhos-Nergis, N. Cimpoesu, M.A. Bernevig-Sava, S.L.
       Toma, D.C. Darabont, C. Baciu, Study on the anticorrosive phosphated steel
       carabiners used at personal protective equipment, Quality-Access to Success 20(1)
       (2019) 71-76.

[11]   D.P. Burduhos-Nergis, C. Nejneru, R. Cimpoesu, A.M. Cazac, C. Baciu, D.C.
       Darabont, C. Bejinariu, Analysis of Chemically Deposited Phosphate Layer on the
       Carabiners Steel Surface Used at Personal Protective Equipments, Quality-Access
       to Success 20(1) (2019) 77-82.

[12]   A.V. Sandu, C. Bejinariu, G. Nemtoi, I.G. Sandu, P. Vizureanu, I. Ionita, C. Baciu,
       New anticorrosion layers obtained by chemical phosphatation, REVISTA DE
       CHIMIE, 64(8) (2013) 825-827.

[13]   A.V. Sandu, C. Coddet, C. Bejinariu, Study on the chemical deposition on steeel
       of zinc phosphate with other metallic cations and hexamethilen tetramine. I.
       Preparation and structural and chemical characterization, Journal of
       Optoelectronics and Advanced Materials, 14(7-8) (2012) 699 - 703.

[14]   P. Lazar, C. Bejinariu, A.V. Sandu, A.M. Cazac, O. Corbu, M.C. Perju, I.G.
       Sandu, Corrosion Evaluation of Some Phosphated Thin Layers on Reinforcing

Steel, IOP Conference Series: Materials Science and Engineering, 209(1) (2017) 012025. https://doi.org/10.1088/1757-899X/209/1/012025

[15]   D.P. Burduhos-Nergis, A.V. Sandu, D.D. Burduhos-Nergis, D.C. Darabont, R.I. Comaneci, C. Bejinariu, Shock Resistance Improvement of Carbon Steel Carabiners Used at PPE. MATEC Web of Conferences, 290 (2019) 12004. https://doi.org/10.1051/matecconf/201929012004

[16]   A.V. Sandu, C. Bejinariu, I.G. Sandu, M.M.A.B. Abdullah, Modern Technologies of Thin Films Deposition. Chemical Phosphatation, Material Research Forum, USA (ISBN 978-1-945291-90-6) (2018) 149.

[17]   A.V. Sandu, C. Coddet, C. Bejinariu, A Comparative Study on Surface Structure of Thin Zinc Phosphates Layers Obtained Using Different Deposition Procedures on Steel, REVISTA DE CHIMIE, 63(4) (2012) 401-406.

[18]   A.V. Sandu, A. Ciomaga, G. Nemtoi, C. Bejinariu, I. Sandu, SEM-EDX and microFTIR studies on evaluation of protection capacity of some thin phosphate layers, Microscopy Research and Technique, 75(12) (2012) 1711-1716. https://doi.org/10.1002/jemt.22120

CHAPTER 6

Corrosion Resistance of the Deposited Layers

As carabiners come into contact with various corrosive substances, one of the main properties that the material used in the manufacture of carabiners must possess is corrosion resistance. Given that the main goal of this research is to improve the corrosion resistance properties of carbon steel carabiners, three phosphate layers with different compositions were deposited on the surface of the material, according to Tables 3.9, 3.10 and 3.11, which are analyzed in terms of corrosion resistance properties.

6.1 Determination of corrosion potential and instantaneous corrosion rate

Corrosion potential (E_{cor}) is a measure of the corrosion tendency of an alloy immersed in an electrolytic medium. It was determined indirectly, by linear polarization curves, using the Evans diagram. This diagram shows the current density logarithm depending on the electrode potential within a $\pm$ 50÷60 mV overpotential range around the corrosion potential. The value of the corrosion potential, $E_{cor} \equiv E(I = 0)$,, is equal to the coordinate point, on the potential axis, of the intersection between the linear section of the anodic branch and that of the cathodic branch of the polarization curve [1-5].

Instantaneous corrosion rate is the rate of corrosion of a metal or alloy immersed in an electrolytic medium to which no potential is applied. In order to determine the corrosion current, at the corrosion potential of the metal or alloy, from the linear polarization curve obtained for relatively low overvoltages, the corrosion rate was determined by the polarization resistance method [6-8]. The anodic polarization curves were recorded using a PGP 201 potentiometer (Radiometer Analytical SAS - France), while the VoltaMaster 4 software was used for experimental data collection and processing.

The anodic polarization curves were recorded with a potential scanning speed of 0.5 mV/s over a potential range of -200 mV to +300 mV compared to the value of the open circuit potential.

In all the studied cases, when interpreting linear polarization data using the VoltaMaster 4 software, a calculation area of 120 mV ($\pm$ 60 mV around the corrosion potential) and a linearity section of 30 mV were considered. In this software, the density of the instantaneous corrosion current (J_{cor}) is calculated using ratio (eq. 6.1), while the corrosion rate, expressed as penetration rate (v_p), was calculated using ratio (eq. 6.2) [3, 9-13].

$$J_{cor} = \frac{b_a \cdot b_c}{2303(b_a + b_c) \cdot R_p}; \ (mA / cm^2) \qquad (6.1)$$

$$v_p = 3{,}27 \cdot \left(\frac{\overline{A}}{z}\right) \cdot \frac{J_{corr}}{\rho}; \ (mm / an) \qquad (6.2)$$

where: - ba and bc are the linear section of the anodic branch and of the cathodic branch in diagram E = f (log J), respectively;

- $R_p = (dE/dj)E_{cor}$ polarization resistance (expressed in ohm·cm²);

- A/z is the electrochemical equivalent of the corroding metal (iron, in our case), as this is the main component of the alloy: A(Fe) = 55.85 g/mol and z = 2);

- ρ – density (for Fe, ρ = 7.5 g/cm³).

Appendix 1 shows the linear polarization curves in semi-logarithmic coordinates (Evans diagram) for the samples studied in the three corrosion media (rainwater, Black Sea water and fire extinguishing solution, respectively).

The experimental results for the samples studied in rainwater (APL) are summarized in Table 6.1.

Table 6.1 Parameters of the instantaneous corrosion process for the samples studied in rainwater.

Sample	C45	I-Zn	II-Zn/Fe	III-Mn	OFU	OFV
E(I=0), mV	-686	-420	-326	-465	passivated	passivated
R_p, kΩ·cm²	11.17	19.40	16.67	14.79	-	-
j_{cor}, μA/cm²	3.15	1.89	2.49	2.38	-	-
v_{cor}, μm/year	38.30	22.97	30.26	28.94	-	-
b_a, mV/decade	196	218	163	154	-	-
b_c, mV/decade	-283	-234	-410	-263	-	-

As shown in Table 6.1, corrosion potentials range within the negative domain. With the exception of the corrosion potential specific to sample *C*45, their absolute value is relatively low, i.e. they are shifted towards the positive domain. Thus, it can be observed that rainwater is not an aggressive agent for the studied materials.

As far as samples *OFU* and *OFV* are concerned, the values of the corrosion currents are of the order of pico amperes and cannot be properly assessed with the device used. This behavior is due to the increase in corrosion resistance of the working electrode surface due to the lubricant in the pores of the phosphate or paint coat. Therefore, the low value of the corrosion current is due to the low conductivity of rainwater and the high polarization resistance of the coating. The polarization resistance of the other samples ranges between 10 kΩ·cm^2 and 20 kΩ·cm^2. The lowest value is determined in the base material, while the polarization resistance of phosphate-coated samples differs depending on the phosphating solution used. Therefore, in terms of corrosion resistance, the analyzed samples rank as follows:

$$C45 < III - Mn < II - Zn/Fe < I - Zn <<< OFU, OFV;$$

while in terms of corrosion current density, the samples rank as follows:

$$C45 > II - Zn/Fe > III - Mn > I - Zn.$$

Also, the corrosion rates have low values (20÷30 μm/year) and hence the integrity of the samples is not affected.

In all the samples that we analyzed, the Tafel slope for the cathodic branch (b$_c$) is higher than that for the anodic branch (b$_a$). This may be due to limitations in the rate of ion tran sfer in solution specific to media with low electrical conductivity. The value of the corrosion rate may be influenced by the Tafel slope, since the current density calculation equation is dependent on it. Therefore, the inversion in the rank of current densities compared to the polarization resistance rank is due to the ratio between the absolute values of the Tafel constants (b$_c$/b$_a$), which are as follows: 1.5 for $C45$, 1.07 for $I - Zn$, 2.5 for $II\text{-}Zn/Fe$ and 1.7 for $III - Mn$.

The experimental results obtained for the samples studied in Black Sea water (AMN) are shown in Table 6.2.

Table 6.2 Parameters of the instantaneous corrosion process for the samples studied in Black Sea Water.

Sample	C45	I-Zn	II-Zn/Fe	III-Mn	OPS	PPS
E(I=0), mV	-636	-748	-578	-766	-581	-163
R$_p$, kΩ·cm^2	1.83	2.74	2.04	0.644	4.11	0.715
j$_{cor}$, μA/cm^2	18.41	15.94	15.97	51.26	9.38	57.91
v$_{cor}$, μm/year	223.9	193.8	194.1	623.5	114.0	704.4
b$_a$, mV/decade	193	393	110	217	223	248
b$_c$, mV/decade	-210	-211	-279	-184	-275	-551

In the case of this corrosive medium, the corrosion potential of the $III - Mn$ sample is shifted to negative values, which indicates an advanced corrosion tendency. Nevertheless, the corrosion tendency of phosphate-coated zinc samples and of lubricant- or paint-coated samples is much lower.

These tendencies are quantitatively confirmed by the values of the polarization resistances, which range between 0.6 kΩ·cm^2 and 4.11 kΩ·cm^2, which are by approximately one order of magnitude smaller than in the case of rainwater corrosion (APL). However, no connection can be set between the corrosion potential values and the polarization resistance values. The largest discrepancy between these is identified in the case of the phosphate-coated and painted steel sample (OFV). It has the lowest absolute corrosion potential value (-163 mV), as well as a polarization resistance of 0.715 kΩ·cm^2. However, the corrosion rate is very high due to the fact that seawater quickly damages the paint used.

The $I - Zn$ sample also show high polarization resistance in this case, due to the accumulation of ionic species involved in the corrosion process in the phosphate coat, which is more compact than that of the $II - Zn/Fe$ and $III - Mn$ samples.

The lubrication of the phosphate-coated sample immersed in the first solution (OFU) improves corrosion resistance. Thus, polarization resistance increases significantly ($R_{p,OFU}/R_{p,I-Zn}$=1.5), while the corrosion rate decrease is about twofold (v_{I-Zn}/v_{OFU} =1.69).

The $III - Mn$ sample has a very low polarization resistance and, at the same time, a high corrosion rate, more than three times higher than that of the phosphate-coated zinc material (v_{III-Mn}/v_{I-Zn}= 3.21). Such behavior may be accounted for by the overlap of galvanic corrosion over direct corrosion. The occurrence of galvanic corrosion may be correlated with the high porosity of the coat (as one may see in Fig. 4.5), as well as with the high content of iron and nickel ions. Thus, micropyles occur on the surface of the alloy between two different materials joined by a strong electrolytic medium.

As far as the fire extinguishing agent (SSI) is concerned, the tested samples showed the following experimental values (Table 6.3).

Table 6.3 *Parameters of the instantaneous corrosion process for the samples studied in fire extinguishing solution.*

Sample	C45	I-Zn	II-Zn/Fe	III-Mn	OPS	PPS
E(I=0), mV	-605	-743	-748	-749	-752	-655
R_p, kΩ·cm^2	0.192	0.372	0.229	0.075	0.566	1.58
j_{cor}, µA/cm^2	146.3	183.4	107.9	355.4	42.49	20.02
v_{cor}, µm/year	177.9	223.0	131.2	432.3	516.8	243.4
b_a, mV/decade	150	53	76	112	52	186
b_c, mV/decade	-195	-481	-321	-289	-271	-213

A comparative study of the aggressiveness of the three media reveals that the absolute corrosion potential values of the fire extinguishing agent are usually higher. Therefore, the fire extinguishing agent is the most aggressive corrosion medium of the three that we analyzed.

Again, the manganese phosphate coat has lower corrosion resistance and the samples rank as follows in terms of polarization resistance:

$$I - Zn > II - Zn/Fe > C45 > III - Mn$$

However, in terms of instantaneous corrosion current density for $C45$ and phosphate-coated samples, and in terms of corrosion rate, the samples rank as follows:

$$II - Zn/Fe < C45 < I - Zn < III - Mn$$

The inversions of the current density values compared to the polarization resistance values are attributed to the auxiliary electrode processes that lead to large differences between the Tafel slopes. Some auxiliary electrode processes are concentration polarization (change in the concentration of active species in the immediate vicinity of the alloy surface, reduction of oxygen dissolved in the solution, mass transfer type - migration, diffusion, convection), ohmic polarization (which occurs when electrolyte resistance is high), hydrogen reduction. All these processes change the active species transfer rate to the electrode (either to the cathode or to the anode).

As in the case of seawater corrosion, the $III - Mn$ sample has the lowest corrosion resistance in the fire extinguishing agent, its corrosion rate being two or even three times higher than that of phosphate-coated zinc samples. The corrosion of this sample in this solution is about seven times higher than in seawater ($v_{SSI}/v_{AMN} \approx 6.9$) and about a hundred times higher than in rainwater ($v_{SSI}/v_{APL} \approx 149.32$).

Lubricant coating and painting of phosphate-coated samples provides better corrosion protection when immersed in SSI, the lowest corrosion rate being achieved by painting the samples. Unlike seawater, fire extinguishing agents do not damage the coat of paint.

6.2 Overpotential behavior. Cyclic potentiodynamic polarization

Cyclic potentiodynamic polarization is one the methods used to characterize corrosion processes. In order to determine cyclic polarization curves, also called cyclic voltammograms, the tested alloy is continuously polarized at a preset potential scanning rate (mV/s) and the current in the circuit is recorded. Potential scanning and current variation recording are automatic, which results in a continuous curve. A high rate of variation of the working electrode potential was used in order to achieve current intensities high enough to cover any accidental system fluctuations, yet low enough to detect all the processes that occur in the solution or on the surface of the electrode. In order for the alloys to be passivated at the beginning of the anodic polarization, the absolute value of the initial potential (-800 mV) higher than the value of the corrosion potential was chosen [3].

Cyclic polarization curve analysis provides information about the type of electrochemical process that takes place on the surface of the tested alloy in an electrolytic medium (such as: generalized corrosion, localized corrosion, passivation, redox processes of solution species), assessment of specific potentials (corrosion potential, penetration potential, re-passivation potential, protection potential [14]. Based on the currents recorded at potentials other than the equilibrium potential E_0, the corrosion rate under load may be calculated (when a higher potential than the corrosion potential is applied to the metal or alloy immersed in the solution).

Depending on the purpose, the experimental data are represented in several ways, the most common of which are $j = f(E)$ and semi-logarithmic representation: $E = f(\log j)$, where E is the potential applied to the alloy, and j – is the total current density at that potential. Fig. 6.1 shows the two types of representation of a cyclic voltammogram using the experimental data collected for the $II - Zn/Fe$ sample in Black Sea water (AMN).

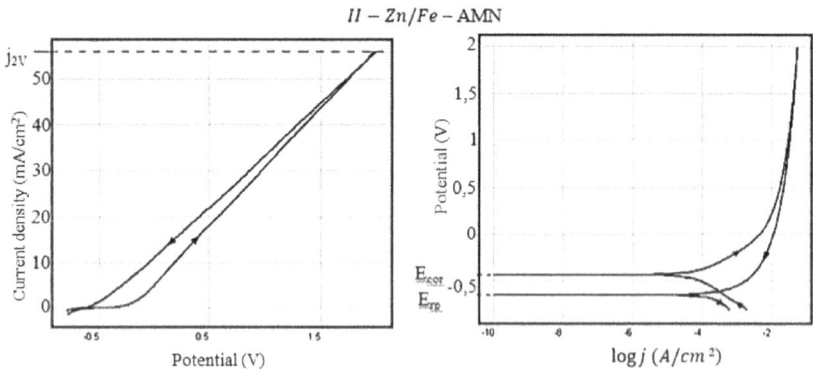

Figure 6.1 Cyclic polarization curve and evaluation method of the parameters of the corrosion process.

The specific values that may be inferred from this curve are: the corrosion potential (E_{cor}),) and the repassivation potential (E_{rp}).

The corrosion potential (E_{cor}) is the potential at which the metal or alloy passes from a passive state, in which no oxidation process takes place, to an active state, when the corrosion of the metal begins [15]. The value of the corrosion potential inferred from this diagram is not equal to the corrosion potential inferred from the linear polarization curve. The difference is due to the fact that the corrosion potential inferred from the linear polarization curve is very close to the equilibrium value, as the potential scanning rate is very low (0.5 mV/s), while in the case of cyclic voltammogram the scanning rate is high (10 mV/s) and the system is not in a state of equilibrium. The corrosion potential corresponds to the potential at which the anodic branch of the cyclic voltammogram changes from negative current density values to positive current density values.

The repassivation potential (E_{rp}) is the potential below which all active corrosion points are repassivated. Below this potential value the metal or alloy is passive (no longer corrodes). The repassivation potential corresponds to the potential at which the cathodic branch of the cyclic voltammogram passes from positive values to negative values of the current density [16].

Both the corrosion potential and the repassivation potential are determined much more accurately from the semi-logarithmic diagram, as shown in Fig. 6.1.

Moreover, in order to be able to draw a comparison between the intensity of the corrosion process in different samples in the same corrosion medium or for the same sample in different corrosion media, the current densities corresponding to the peak overpotential used in this research, i.e. (j_{2v}) were also determined in the cyclic polarization diagrams. This may be more accurately determined in diagram $j = f(E)$ and in Fig. 6.1.

Appendix 2 shows the cyclic voltammograms for the samples analyzed in the three corrosion media, represented in semi-logarithmic coordinates.

The shape of the cyclic voltammograms and the position of the anodic and cathodic branches allow the collection of information on the type of corrosion (generalized corrosion, point corrosion, passivation, etc.).

These curves were used to assess the parameters specific to the processes that take place when a relatively high potential is applied to the alloy. The research is important since this process accelerates the processes that may occur on the surface of the alloy immersed in the corrosion medium and thus the behavior of the alloy if it were immersed for a long time in solution may be predicted.

In the case of voltammograms that in certain potential fields show linear variation of their current density as a function of the potential, the equations of the linear sections on the anodic and/or cathodic branch were also analyzed. Since in some cases a leap occurs on the anodic branch (direct polarization curve, from negative values to positive values) at a particular current density value, the tables only show the linear equations of the cathodic branches.

Cyclic voltammograms were also used to determine the corrosion rate based on the anodic polarization curve drawn at a high potential scanning rate (10 mV/s). The Tafel method (polarization resistance method) applied to the points on the anodic branch of the polarization curve located in the vicinity of the corrosion potential was used for this purpose ($\pm$ 120 mV compared to E_{cor}). Polarization resistance (R_p), corrosion current density (j_{cor}) and corrosion rate (v_{cor}) were thus determined [17-19]. These may be compared with data collected at low potential scanning rate and reveal the extent to which the scanning rate can influence results.

6.2.1   Overpotential behavior of the C45 sample

The cyclic polarization curves drawn for the freshly ground $C45$ sample in the three corrosion media are shown in Fig. 6.2, while the specific overpotential behavior parameters are included in Table 6.4.

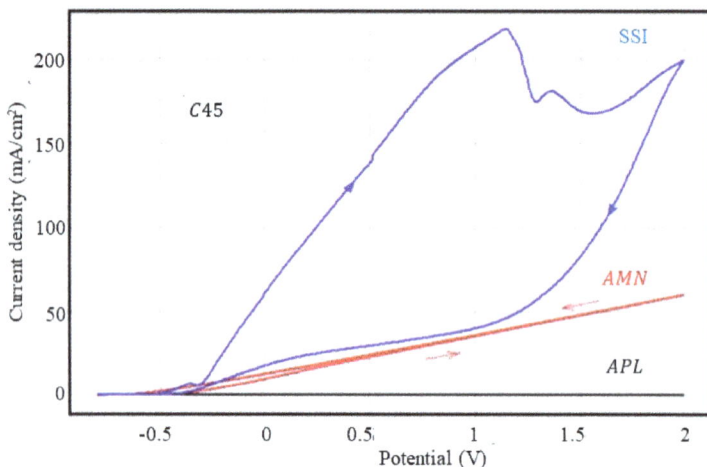

Figure 6.2 Cyclic voltammograms for C45sample.

As one may notice in Fig. 6.2 and Table 6.4, rainwater is not an aggressive corrosion agent, as it does not significantly alter the surface of the alloy, while the fire extinguishing solution is the most aggressive agent. Thus, when the current density value at 2V overpotential is considered as an aggressiveness criterion, we may report that: $j_{2V(SSI)}:j_{2V(AMN)}=3,28$; $j_{2V(SSI)}:j_{2V(APL)}=519$ and $j_{2V(AMN)}:j_{2V(APL)}=159$. According to this criterion, the fire extinguishing agent is 500 times more aggressive than rainwater and more than three times more aggressive than seawater.

Table 6.4 *Stimulated corrosion parameters for the C45 sample in the three corrosive environments.*

Corrosive environment	E_{cor}, mV	E_{rp} mV	$j(mA/cm^2)=a \cdot E(V)+b$		j_{2V} mA/cm²	Tafel parameters at v_s=10 mV/s		
			a	b		R_p kW·cm²	j_{cor} μA/cm²	V_{cor} μm/year
APL	-276	+564	0.254	-0.127	0.381	13.77	2.572	31.28
AMN	-693	-734	23.76	+12.25	60.39	1.56	6.594	80.20
SSI	-653	-674	-	-	198.0	0.962	21.523	261.8

The anodic and cathodic curves are virtually linear within the 0 - 2000 mV potential range, in both rainwater and Black Sea water. The two voltammogram branches are very close, which means that a process of generalized corrosion is taking place.

The anodic and cathodic branches of the voltammogram recorded in the fire extinguishing solution are arched and very far from each other. The anodic branch (direct curve) is above the cathodic curve (reverse curve), which indicates that significant metal passivation occurs during cyclic polarization. This behavior is due to the fact that some of its organic substances are preferentially adsorbed on the surface of the alloy, influencing the repassivation potential. Thus, while in seawater and stormwater the repassivation potential is lower than the corrosion potential, in fire extinguishing solution the repassivation potential is close to the corrosion potential.

Table 6.5 shows a comparison between the corrosion rates determined at low potential scanning rate and those determined at high scanning rate. Surprisingly enough, only in rainwater measurements the corrosion rate (determined at a potential scanning rate of 10 mV/s) is very close to the rate determined on the linear polarization curve drawn at a scanning rate of 0.5 mV/s. Corrosion rate in sea water is much lower at high scanning rates than at low rates.

6.2.2 Overpotential behavior of the I-Zn sample

Fig. 6.3 and Table 6.6 show the cyclic polarization measurements for the phosphate-coated steel sample immersed in the first solution ($I - Zn$).

The aggressiveness of the corrosion medium increases for this sample in the same order as for sample $C45: APL < AMN < SSI$, the ratios between the J_{2v} values in the three solutions are as follows: $j_{2v(SSI)}{:}j_{2v(AMN)} = 3{,}65$; $j_{2v(SSI)}{:}j_{2v(APL)} = 624$ and $j_{2v(AMN)}{:}j_{2v(APL)} = 171$, i.e. a little higher than in the C45 sample, but of the same order of magnitude.

Table 6.5 Comparison between corrosion rates (v_{cor} - in µm/year) for C45 sample.

Corrosive environment	Scan rate	
	0.5 mV/s	10 mV/s
Rainwater	38.3	31.3
Black Sea Water	223.9	80.2
Fire extinguishing solution	177.9	261

Table 6.6 Stimulated corrosion parameters for the I-Zn sample in the three corrosive environments.

Corrosive environment	E_{cor}, mV	E_{rp} mV	$j(mA/cm^2)=a \cdot E(V)+b$ a	b	j_{2v} mA/cm²	Tafel parameters at v_s=10 mV/s R_p kW·cm²	j_{cor} µA/cm²	V_{cor} µm/year
APL	+712	+996	0.172	-0.179	0.168	30.60	1.352	16.44
AMN	-392	-764	-	-	28.74	5.63	4.754	57.83
SSI	-475	-520	-	-	104.9	1.10	28.55	347.33

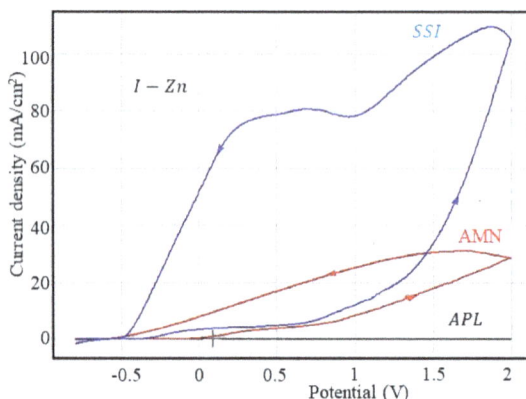

Figure 6.3 Cyclic voltammograms for I-Zn sample.

Phosphate-coating of the C45 steel sample immersed in the first solution increases corrosion resistance in rainwater (the corrosion rate decreases by about two times) and in Black Sea water (v_{cor} decreases by 1.4 times), yet, surprisingly enough, it increases corrosion rate in the fire extinguishing solution, as shown in Table 6.7.

Table 6.5 Comparison between corrosion rates (v_{cor} - in µm/year) for I-Zn sample.

Corrosive environment	Scan rate	
	0.5 mV/s	10 mV/s
Rainwater	22.97	16.44
Black Sea Water	193.8	57.83
Fire extinguishing solution	223	347

The cathodic branch is located above the anodic branch in Black Sea water and in fire extinguishing solution, which is an indication of an increase in the corrosion rate following the anodic process (the return current density (on the cathodic branch) is higher than on the direct branch (anodic branch) at the same potential). The increase in corrosion after anodic polarization is very high in SSI.

6.2.3 *Overpotential behavior of the II-Zn/Fe sample*

The cyclic polarization curves determined for $II-Zn/Fe$ in the corrosion media used in this research are shown in Fig. 6.4, while the specific overpotential behavior parameters are synthetically described in Table 6.8.

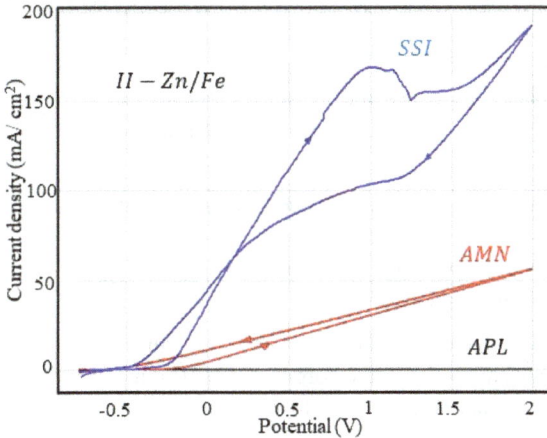

Figure 6.4 Cyclic voltammograms for II-Zn/Fe sample.

Table 6.8 Stimulated corrosion parameters for the II-Zn/Fe sample in the three corrosive environments.

Corrosive environment	E_{cor}, mV	E_{rp} mV	$j(mA/cm^2)=a \cdot E(V)+b$		j_{2V} mA/cm²	Tafel parameters at $v_s=10$ mV/s		
			a	b		R_p kW·cm²	j_{cor} μA/cm²	V_{cor} μm/year
APL	+88	+752	0.24	-0.17	0.321	10.63	2.94	35.84
AMN	-465	-631	22.68	+10.93	56.04	0.66	45.28	550
SSI	-533	-686	-	-	193.1	0.15	189.60	2306

Just like in the $C45$ and $I-Zn$ samples, rainwater is a hardly aggressive liquid medium, while the fire extinguishing agent is very aggressive, although seawater is considered to be an aggressive corrosion medium due to its high chlorine concentration.

Despite the considerable differences between the corrosion rates of $C45$ and $II-ZN/Fe$, the peak values at 2V are very close in all three corrosion media: the $j_{2V(II-ZN/Fe)}:j_{2V(C45)}$ ratio is 0.86 in rainwater, 0.92 in sea water and 0.97 in fire extinguishing agent. This behavior could be accounted for by the corrosion rate decrease during the anodic polarization of the phosphate-coated sample. The corrosion rates listed in these tables are determined at the corrosion potential, even in the initial moments of polarization, when the pores of the phosphate film are free. As the potential increases and the reaction evolves, the reaction products formed on the surface (most likely Fe_2O_3 or Fe_3O_4) cover the pores and thereby reduce the flow of the corrosive agent towards the metal.

The two branches of the cyclic polarization curve are perfectly linear in seawater; the current density (and thus the corrosion rate) is directly proportional to the potential applied to the alloy. The cathodic (return) branch is located slightly above the anodic branch, which means that the phosphate-coated alloy does not passivate but, on the contrary, the corrosion rate increases, probably due to the increase of the roughness of the alloy surface due to corrosion.

In the fire extinguishing solution, the cathodic branch is located below the anodic branch, which indicates a rather significant passivation process. In this case, passivation occurs by clogging the pores with corrosion products, as well as by adsorption of organic components from the solution.

Unlike the $C45$ and $I-Zn$ samples, in this case, the potential scanning rate has a different (inverse) influence on the corrosion rate determined by the Tafel method, the rate determined by polarization with high scanning rate is much higher than at low scanning rate (Table 6.9).

Table 6.9 Comparison between corrosion rates (v_{cor} - in μm/year) for II-Zn/Fe sample.

Corrosive environment	Scan rate	
	0.5 mV/s	10 mV/s
Rainwater	30.26	35.84
Black Sea Water	194.10	550.00
Fire extinguishing solution	131.20	2306.00

6.2.4   Overpotential behavior of the III-Mn sample

The cyclic polarization curves determined for III-Mn in the corrosion media used in this research are shown in Fig. 6.5, while the specific overpotential behavior parameters are synthetically described in Table 6.10.

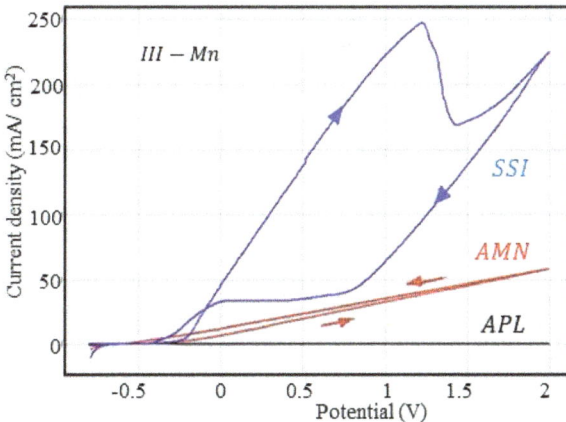

Figure 6.5 Cyclic voltammograms for III-Mn sample.

Table 6.10 Stimulated corrosion parameters for the III-Mn sample in the three corrosive environments.

Corrosive environment	E_{cor}, mV	E_{rp} mV	$j(mA/cm^2)=a \cdot E(V)+b$		j_{2V} mA/cm²	Tafel parameters at v_s=10 mV/s		
			a	b		R_p kW·cm²	j_{cor} μA/cm²	V_{cor} μm/year
APL	-178	+454	0,26	-012	0.42	5.27	7.46	90.78
AMN	-558	-611	23.27	+11.83	58.20	0.20	176.60	214.80
SSI	-558	-520	-	-	224.00	0.16	182.30	221.70

The shapes of the cyclic voltammograms in the three corrosion media are similar to those of the other samples. In terms of peak current density (j_{2v}), in the phosphate-coated sample series, III-Mn is the sample with the lowest corrosion resistance, the ranking according to corrosion resistance being as follows III-Mn<II-Zn/Fe<I-Zn.

The same ranking is revealed by the analysis of the corrosion rate values determined by the polarization resistance method at a potential scanning rate of 10 mV/s. An exception to this rule is found in the case of measurements made in a fire extinguishing solution, where an inversion between II-Zn/Fe and III-Mn occurs, probably due to higher passivation in III-Mn (related to substance adsorption from SSI).

The anodic and cathodic branches of the voltammogram are linear in rainwater and seawater. The location of the cathodic curve above the anodic branch indicates generalized corrosion (even over the entire surface in contact with the solution), with a slight increase in the corrosion rate after traversing the anodic branch.

Just like in II-Zn/Fe (yet unlike in I-Zn), the cyclic voltammogram is typical of an alloy that is passivated, the cathodic branch is located below the anodic branch, and the distance between them is considerable (passivation is quite significant).

The corrosion rate at 10 mV/s is much higher than the one determined when the scanning rate is 0.5 mV/s, as shown in Table 6.11.

Table 6.11 Comparison between corrosion rates (v_{cor} - in µm/year) for III-Mn sample.

Corrosive environment	Scan rate	
	0.5 mV/s	10 mV/s
Rainwater	28.94	90.78
Black Sea Water	623.50	2148.00
Fire extinguishing solution	432.30	2217.00

The SSI measurement is once again an exception, probably also due to its complex composition and likely surface adsorption.

6.2.5 *Overpotential behavior of the OFU sample*

The cyclic polarization curves determined for the phosphate-coated $C45$ sample immersed in the first solution and impregnated in lubricant are shown in Fig. 6.6, while the specific overpotential behavior parameters are listed in Table 6.12.

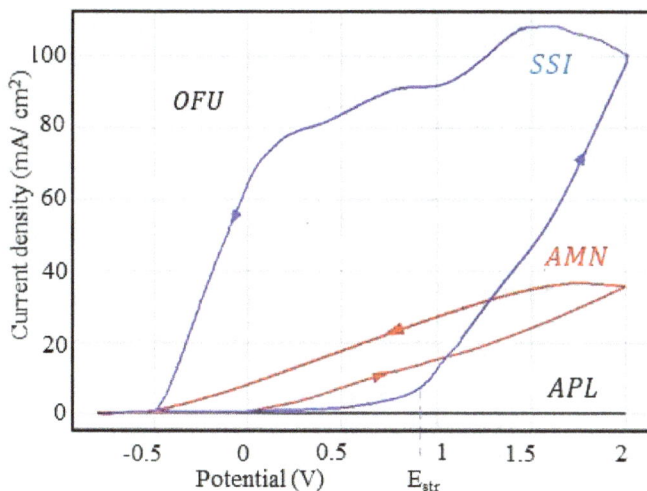

Figure 6.6 Cyclic voltammograms for OFU sample.

Table 6.12 Stimulated corrosion parameters for the OFU sample in the three corrosive environments.

| Corrosive environment | Ecor, mV | Erp mV | j(mA/cm²)=a·E(V)+b | | j2v mA/cm² | Tafel parameters at vs=10 mV/s | | |
			a	b		Rp kW·cm²	jcor µA/cm²	Vcor µm/year
APL	+321	+852	0.18	-0.15	0.20	8.90	2.19	26.75
AMN	-458	-758	-	-	35.47	7.13	4.53	51.17
SSI	-576	-502	-	-	104.20	1.02	30.28	376.00

The shapes of the cyclic voltammograms in the three corrosion media are similar to those obtained for the other samples; linear variations on the anodic and cathodic branches within the 0 - 2000 mV potential range in rainwater and in Black Sea water (except for sample $I - Zn$).

In rainwater, the j2v and corrosion rate are very low and hardly significant for the occurrence of dangerous material corrosion.

In the Black Sea water, the cathodic branch of the voltammogram is located above the quite spaced anodic curve, which indicates that the sample surface was affected by the corrosion produced during anodic polarization, thus leading to an increase in the corrosion rate on the cathodic branch (the current density on the cathodic branch is higher than on the anodic branch at similar potential). In seawater, the corrosion rate of this sample has the lowest value both as compared to $C45$ and as compared to the phosphate-coated samples immersed in the three solutions (yet very close to the $I - Zn$ rate – which was also used in lubricant coating).

In the fire extinguishing solution, the lubricated sample exhibits a sharp decline in corrosion resistance. On the anodic curve in the field of low potentials the corrosion rate increases with potential increase, up to Estr, after which it increases sharply with potential increase. Estr is called penetration potential and is due to the penetration of the protective coat (probably the phosphate coat) but also to the simultaneous lubricant removal. The behavior of this sample in a fire extinguishing solution is very similar to the behavior of the $I - Zn$ sample, a material that was used for lubricant coating. The other two phosphate-coated samples behave completely differently when polarized in a fire extinguishing solution; both samples are easily passivated after passing through the anodic branch; in these cases the phosphate coat is probably not destroyed by anodic polarization, and passivation occurs by adsorption of some products from the fire extinguishing solution.

As shown in Table 6.13, no correlation may be established between the corrosion rates measured at the two potential scanning rates.

Table 6.13 Comparison between corrosion rates (v_{cor} - in µm/year) for OFU sample.

Corrosive environment	Scan rate	
	0.5 mV/s	10 mV/s
Rainwater	-	26.80
Black Sea Water	114	51.17
Fire extinguishing solution	517	376.00

6.2.6   Overpotential behavior of the OFV sample

The cyclic polarization curves obtained for OFV are shown in Fig. 6.7, while the specific overpotential behavior parameters are listed in Table 6.14.

The cyclic polarization curve for the painted sample immersed in rainwater could not be determined because the coat of paint is compact and does not conduct electricity or because rainwater is not an aggressive agent and does not damage the coat of paint.

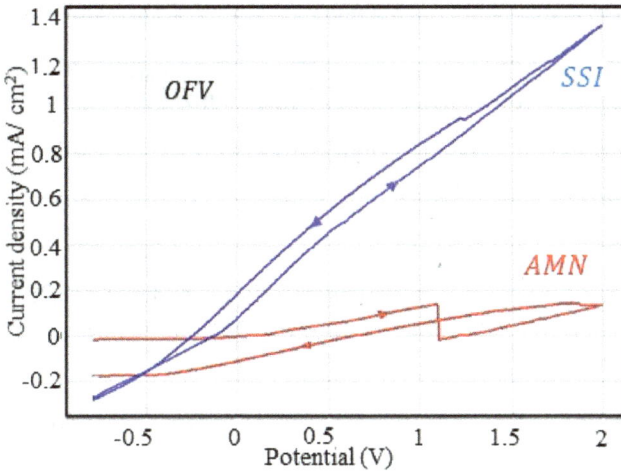

Figure 6.7 Cyclic voltammograms for OFV sample.

Table 6.14 Stimulated corrosion parameters for the OFV sample in the three corrosive environments.

Corrosive environment	E_{cor}, mV	E_{rp} mV	$j(mA/cm^2)=a \cdot E(V)+b$		j_{2v} mA/cm²	Tafel parameters at $v_s=10$ mV/s		
			a	b		R_p kW·cm²	j_{cor} μA/cm²	V_{cor} μm/year
APL	-	-	-	-	-	-	-	-
AMN	+84	+664	-	-	0.13	12.20	2.78	33.92
SSI	-149	-243	0.52	+0.30	1.35	2.01	17.75	216.00

Seawater and the fire extinguishing agent affect the quality of the coat of paint, by widening the micropores that already exist in the coating layer, thus allowing the corrosive agent to penetrate the phosphate coat and the alloy. However, the coat of paint provides better corrosion resistance than phosphate films even for the lubricated sample.

Table 6.15 shows a comparison between the corrosion rates (v_{cor}) and peak overpotential current densities (j_{2v}) of the *OFV* sample and those of the other samples considered in our research, immersed in seawater and fire extinguishing agent.

Table 6.15 Comparison between corrosion rates and current densities between OFV and other samples.

Parameters ratio	Corrosive environment	Sample (S)				
		C45	I-Zn	II-Zn/Fe	III-Mn	OFU
$(v_{cor})_S/(v_{cor})_{OFV}$	AMN	2.36	1.70	16.21	6.33	1.51
	SSI	1.21	1.61	10.67	9.94	1.74
$(j_{2V})_S/(j_{2V})_{OFV}$	AMN	454	216	421	437	266
	SSI	146	77.40	142	165	76

Corrosion potentials are positive for seawater and considerably shifted towards the positive range for the fire extinguishing agent, which is also an indication of low corrosion tendency.

6.3 Analysis of the structure of the alloy/corrosion medium interface by Electrochemical Impedance Spectroscopy

When a metal or alloy with a freshly ground surface or coated by various films is immersed in an electrolytic medium, a series of interactions occur that alter the metal/electrolyte interface.

The electrochemical corrosion reaction of the metal or alloy components, a process that is generally an oxidation reaction, is the main interaction. This allows a layer of positive ions to gather on the surface of the metal, and a layer of negative ions to gather in front of it from the solution, thus forming the electrical double layer (EDL). This layer prevents the further development of the corrosion reaction and influences the reaction rate, thus representing the kinetic factor of the reaction. The reaction products may be soluble and pass into the corrosion medium or insoluble and may pass into the solution, be deposited on the surface of the alloy without being too adherent or form a compact film strongly adhering to the metal. When the metal is coated with a porous layer, insoluble reaction products may partially or completely cover the pores [20].

Another effect of the alloy/corrosion medium interaction may be the partial degradation of the coating layer, by widening the pores or by detaching it from the alloy, changing the thickness by absorbing the liquid in the film, or adsorbing some ions or molecules from the corrosion medium [21].

In some cases, the corrosion process rate is influenced by the coats deposited on the metal surface (previously deposited, adsorbed from the solution or resulting from the chemical reaction). The influence on the rate of the process is due to the diffusion of ions or molecules involved in the oxidation process. In these cases, the corrosion process is influenced both by the kinetic factor (by EDL) and by the diffusion factor [22].

Sometimes, only adsorption processes of some ions or molecules in the solution take place at the metal/solution interface, yet they do not influence the corrosion process.

Among other electrochemical methods for studying corrosion processes, Electrochemical Impedance Spectroscopy (EIS) is particularly important, as it allows the characterization of electrochemical systems and the determination of the contributions of electrode processes and electrolytic processes to the evolution of these systems. EIS data may be used to establish an equivalent electrical circuit that is directly correlated with the metal/coat and coat/solution interfaces and the phenomena that occur in the passive coat [14, 23]. EIS measurements allow the collection of information about the mechanism of the various processes, and the definition of a theoretical transfer function and the development of a passive coat growth model [24-29].

The electrochemical impedance spectroscopy method is particularly suitable for the study of surfaces with high electrical or electrochemical impedance, which makes it particularly convenient for the determination of damages to metals coated with organic layers (paints) with high electrical resistance. Furthermore, the method allows not only the quantitative determination of the protective features of the coating, but also the collection of information on its mechanism of protection and degradation. Another advantage of this method is the use of a non-destructive technique (by applying a very small alternative electrical signal to the sample), so that repeated determinations are possible on the same sample without altering its properties [30, 31].

6.3.1   Theoretical considerations

In electrochemical impedance measurements, a perturbation is applied to a corrodible metal in the form of a sinusoidal potential, E(t), and the system reaction is recorded, I(t).

where: E_m – peak potential;

ω – angular frequency ($\omega = 2\pi f$, f – frequency in Hz);

I_m – peak current;

φ – the phase angle (phase shift) between current and potential.

The applied signal is a very small potential for a linear dependence between the potential and the current. The measuring equipment records the current-time and potential-time curves as a function of frequency, based on which the impedance spectrum is determined.

In alternating current the impedance is the equivalent of resistance in direct current. Ohm's law is used for the current (I)/voltage (E) ratio, in direct current: $E = I \cdot R$ (R being the resistance), while the $E = I \cdot Z$ ratio (Z being circuit impedance) is used in alternating current. Unlike resistance, the impedance of a circuit depends on the frequency of the applied signal: $Z(\omega) = E(\omega)/I(\omega)$, where E and I are expressed by ratios (6.3) and (6.4).

The reverse of impedance is called admittance and is usually marked Y ($Y = 1/Z = I/E$). Electrochemical impedance is a fundamental feature of the electrochemical system that it represents. Knowing the frequency dependence of impedance in a corrodible system makes it possible to define an appropriate equivalent electrical circuit to describe this system. Such an equivalent electrical circuit consists of resistors, capacitors, inductors (coils) and other more complex magnitudes [32].

An electrical circuit element corresponds to each physical or chemical property of the studied system, the most important of which are:

R_s - resistance of the solution between the reference electrode and the measuring electrode (studied metal);

C_{dl} - capacitance of the electrical double layer at the electrode/electrolyte interface;

R_p - polarization resistance - resistance that opposes the change of the equilibrium potential (open circuit potential) due to the current passing through the metal/solution interface, when the electrode corrodes at this potential, the process being controlled by the equilibrium between the anodic reaction and the cathodic reaction;

R_{ct} - resistance due to charge transfer (ct-charge transfer) - is similar to Rp but it occurs when the process is kinetically controlled by a single reaction (no mixed potential occurs);

W - Warburg impedance is an impedance that occurs when a diffusion process occurs at the solution/electrode interface. This parameter is dependent on the frequency of the disturbing potential;

C_{ext} - the capacitance of the coating layer occurs when two conductive planes are separated by a non-conductive dielectric (the case of metals coated with polymers or organic paints);

Q - constant phase element - which is introduced when in EIS experiments the capacitors do not behave ideally. The constant phase element is a circuit element which, depending on the system, may be assimilated with either a capacitor or a resistor or an imperfect capacitor (when the surface coat is irregular). The transfer functions corresponding to these circuit elements are shown in Table 6.16 below:

Table 6.16 List of linear circuit elements and corresponding transfer functions.

The circuit element	Symbol	Impedance	Admission	Dominant process
Resistor	R	$Z(\omega) = R$	$1/R$	Controlled
Capacitor	C	$Z(\omega) = -1/j\omega C$	$j\omega C$	reactions of the load transfer
Inductor (coil)	L	$Z(\omega) = j\omega L$	$-1/j\omega L$	Adsorption of ions or molecules on the surface
Warburg impedance	W	$Z(\omega) = \sigma \dfrac{1-j}{\omega^{1/2}}$	$\chi \cdot (j\omega)^{1/2}$	Diffusion (mass transfer); σ - diffusion coefficient
Constant phase element	Q	$Z(\omega) = \dfrac{1}{Q(j\omega)^n}$	$Q(j\omega)^n$	$n = 0 \rightarrow Q = 1/R$ $n = 1 \rightarrow Q = C$ $n = -1 \rightarrow Q = 1/L$

The Q constant phase element is introduced in the case of dielectric losses, when dielectric losses occur at the metal/electrolyte interface and dielectric permittivity becomes a function of frequency and is represented by a complex number. Therefore, the electrical double layer or the capacitance of a passive coat can no longer be represented by an ideal capacitor, but by a Constant Phase Element (CPE), which better explains the deviation of Nyquist diagrams from the ideal behavior (a sermicircle on the abscissa axis, Z_r) due to changes in capacitances with frequency. From the mathematical point of view, the impedance of a constant phase element may be expressed in several ways [33-36], yet the most common is the one suggested by Zoltowski [37]:

Where: Q-is a constant proportional to the active area; $<Q> = \Omega^{-1}$ sn/cm$^2 \equiv$ S·sn/cm^2, ω - angular frequency ($\omega = 2\pi f$, f-frequency of alternating current), j - is the imaginary number; $j = (-1)^{1/2}$. The frequency exponent may range from (-1) to (+1): CPE is a pure capacitor when n=1, a pure resistor when n=0, an inductor when n=-1 and a Warburg impedance when n=½.

A consequence of this simple relationship is the fact that the phase angle of CPE is independent of frequency and amounts to $(90°)^n$, which is also the reason why it is called constant phase element [38].

Warburg impedance is introduced when the diffusion of active species from the solution to the metal surface and vice versa is high enough and able to influence the corrosion reaction rate. Diffusion impedance may be represented, in analogy to the CPE impedance, by the ratio:

The equation is applicable to linear diffusion, the unit of measurement of the W factor being: $<Q> = \Omega^{-1}\ sn^{\frac{1}{2}}/cm^2 \equiv S \cdot s^{\frac{1}{2}}/cm^2$. Diffusion impedance is generally linked in series with resistance due to charge transfer (R_{ct}). Thus, charge transfer to and from the surface of the alloy is controlled both kinetically (by R_{ct} and C_{dl}), and by the diffusion through the coat of products deposited on the surface.

The 'ZSimWin' program (Princeton Applied Research, author: Bruno Yeum - EChem Software An Arbor, Michigan, USA) was used in this doctoral thesis to analyze impedance data. The program uses a wide variety of equivalent electrical circuits to numerically correlate the measured impedance data. The program is able to conduct analysis of very complicated dispersion data by decomposing the complex response into that of simple subcomponents. This approach together with the general procedure of nonlinear least squares correlation allows the development of an equivalent circuit whose simulated response is highly correlated with the measured data. One or more time constants may occur in phase angle dependence on frequency, and these may be used to determine the parameters in the equivalent circuit. The Bode spectrum may reveal the presence of a compact passive film if the phase angle in the high frequency range is close to $90^{\circ 8}$ and if the spectrum has linear sections at intermediate frequencies.

The criteria used to choose the most appropriate equivalent circuit require a minimum number of circuit elements, the χ^2 error to be small enough ($\chi^2 < e^{-3} = 0.05$), and the error associated with each element to be less than 5%. The χ^2 factor is the sum of the squares of the differences between the experimental values of impedance and the values calculated by successive iterations, and it represents the extent to which the experimental data are correlated with the calculated ones.

6.3.2 Experimental results

The condition and evolution of alloy/corrosion medium systems were studied one hour after they have been put in contact. This hour includes the thermostatization time and the time required to reach molecular scale equilibria (wetting, ionic equilibria).

The following alloys were used:

- base material $-C45$,

- phosphate-coated C45 steel samples immersed in the first $Zn_3(PO_4)_2 - I - Zn$ solution [39];

- phosphate-coated C45 steel samples immersed in the second $Zn_3(PO_4)_2 + Fe - II - Zn/Fe$ solution;

- phosphate-coated C45 steel samples immersed in the third $Mn_3(PO_4)_2 + Fe + Ni) - III - Mn$ solution [40];

- $I - Zn$ samples coated with lubricant $- OFU$ [41];

- $I - Zn$ samples coated with paint $- OFV$ [42].

The following corrosion media were used: rainwater – APL, Black Sea water – AMN and fire extinguishing agent – SSI.

Electrochemical Impedance Spectroscopy measurements were carried out using a PGZ 301 potentiometer (Radiometer Analytical SAS - France), while experimental data acquisition was performed by means of the VoltaMaster 4 software. The experimental data were processed using the ZSimpWin program [14], which interprets spectra using Boukamp's fitting process - the least squares method. In order to be able to process with this program the data acquired with the VoltaMaster 4 program, they were converted using the EIS file converter program (Radiometer Analytical S.A.).

Electrochemical impedance spectroscopy determinations were conducted using a three-electrode corrosion cell of the C145/170 type (Radiometer, France), which is a glass cell allowing the free flowing of the liquid corrosion medium. The samples (10mm-diameter flat discs) are mounted in the working electrode by means of Teflon washers, which allow the creation of flat circular surfaces of up to 0.8 cm^2. In the analyzed samples, the surface exposed to the corrosion medium was S = 0.503 cm^2. A flat platinum electrode (S = 0.8 cm^2) was used as auxiliary electrode, and a saturated calomel electrode was used as reference electrode. All potentials were measured in relation to this electrode, yet for simplicity reasons, this was not specified in the tables and in the text. The solutions used were naturally aerated and the working temperature was 25 °C [25].

The working sequence for all measurements was: working potential = open circuit potential, frequency range 104 ÷ 10-2 Hz, sinusoidal potential amplitude = 10 mV. Measurements at frequencies higher than 104 Hz were avoided, as a series of preliminary tests indicated the presence of a negative loop in the very high frequency range in the Nyquist diagram. This behavior is probably due to the very high resistance of the reference electrode [43].

Electrochemical impedance data are represented in the form of two types of diagrams; the Nyquist diagram in which the abscissa represents the real part of the impedance, and the ordinate, the imaginary part of the measured impedance (-ImZ vs. ReZ), and the Bode diagram in which the impedance modulus is represented according to the frequency logarithm and the phase angle according to the frequency logarithm (lZl vs. Frequency, and θ vs. Frequency). The advantage of the Bode diagram is that it includes data for all measured frequencies and that a wide range of impedance values can be represented. Frequency dependence on phase angle indicates that one or more time constants occur, which may be used to determine the value of the parameters in the equivalent circuit. Another significant advantage of such a diagram is the possibility to detect areas that are dominated by resistive elements (R_s and R_p) where a slope equal to zero and areas that are dominated by capacitive elements are observed - for which, ideally, a slope equal to (–1) is obtained. All Nyquist and Bode diagrams for the analyzed systems are shown in Appendix 3.

The processing of experimental data for all 18 alloy/corrosion medium systems allowed the identification of a number of 7 equivalent circuits which may be used to analyze the surface properties of the alloys immersed in the liquid medium for one hour. When choosing these circuits, the lowest χ^2 factor and ε_Z relative impedance measurement error were considered.

The presentation of the equivalent circuits that best indicate the experimental data will be done using Boukamp coding, which is used in the ZSimpWin data interpretation software.

6.3.2.1 R(QR(LR)) equivalent circuit

The equivalent electrical circuit shown in Fig. 6.8 a) best describes only the experimental data for the C45/AMN system. One hour after the sample has been immersed in seawater, the Nyquist diagram (Fig. 6.8 b)) for this sample shows a depressed semicircle in the high frequency range and an inductive loop in the low frequency range. This inductive loop may be attributed to the adsorption of ions, neutral molecules or even particles (insoluble corrosion products) in the solution, but it may also indicate the existence of intermediate reactions related to the adsorption of aggressive chlorine ions, which trigger electrode surface instability [22, 36, 44, 45].

Better data adjustment was achieved by replacing the double layer capacitance with a constant phase element, namely CPE, which expresses the non-ideal behavior of the electrical double layer capacitance (capacitance change with frequency). The equivalent circuit consists of a resistor (R_s) in series with a parallel combination of a constant phase element (CPE), a resistor (R_{ct}) and a series resistor-inductance combination (R_L, L). The values of the parameters that best satisfy the experimental data are shown in Table 6.17.

Table 6.17 Equivalent circuit parameters for C45 after one hour of immersion in Black Sea water.

R_s, $\Omega \cdot cm^2$	CPE		R_{ct}, $\Omega \cdot cm^2$	R_L, $\Omega \cdot cm^2$	L, $H \cdot cm^2$	$10^3 \cdot \chi^2$	ε_z
	Q, Ss^n/cm^2	n					
39.6	$3.418\ 10^{-4}$	0.780	588	1007	496	5.92	6.34
ε_{EC}, % 1.22	5.6	1.59	3.95	5.4	2.87	-	-

ε_z–is the impedance measurement error (in percentages), and ε_{EC} – relative percentage error for circuit elements

The values of the χ^2 parameter and of the percentage errors for Z and for the circuit elements show a satisfactory data adjustment with this circuit.

Charge transfer resistance (R_{ct}) is relatively low, which indicates a high corrosion rate. The surface of the sample becomes rough due to corrosion, which makes the capacitance of the electrical double layer non-ideal, thus requiring the introduction of the constant phase element (CPE). The frequency exponent value deviation (n) from 1 is a measure of the deviation from ideality of the capacitor which represents the capacitance of the electrical double layer. The polarization resistance in this case is calculated by the following ratio: $R_p = (R_{ct} \cdot R_L)/(R_{ct} + R_L)$. No continuous layer was formed on the sample surface, only ions from the solution and/or insoluble corrosion products were adsorbed. Therefore, the R_L-L series grouping was introduced in the equivalent circuit, in parallel with the time constant specific to the double layer (CPE-R_{ct}). The R_L circuit element represents the resistance due to the components adsorbed on the surface and L is an inductor (coil) which in SIE describes the inductive loops in the low frequencies range.

The Bode diagram reveals the existence of a single peak on the $\theta = f(v)$ curve (phase angle depending on the frequency of the applied signal), which implies the existence of a single constant of the relaxation time.

6.3.2.2 R(C(R(QR))) equivalent circuit

The equivalent circuit shown in Fig. 6.9 was most suitable for 'fitting' the data (empirical curve adjustment) collected for the C45/APL system, for all three phosphate-coated samples immersed in seawater ($I - Zn$/AMN, $II - Zn/Fe$/AMN and $III - Mn$/AMN) and for the lubricated sample immersed in rainwater (OFU/APL). The circuit describes a bi-layer model, in which the (R_{ext}, C_{ext}) pair represents the properties of the coating layer, and the (CPE, R_{ct}) pair represents the properties of the electrical double layer, which are responsible for the kinetics of the corrosion process. R_s is the uncompensated resistance of the electrolyte column between the reference electrode and the working electrode.

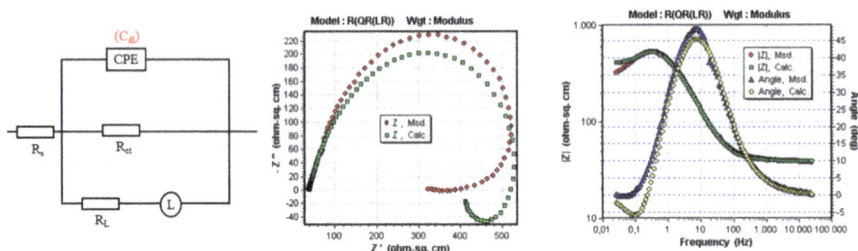

Figure 6.8 Equivalent circuit (a), Nyquist diagram (b) and Bode diagrams (c) for the C45/AMN system after one hour of immersion.

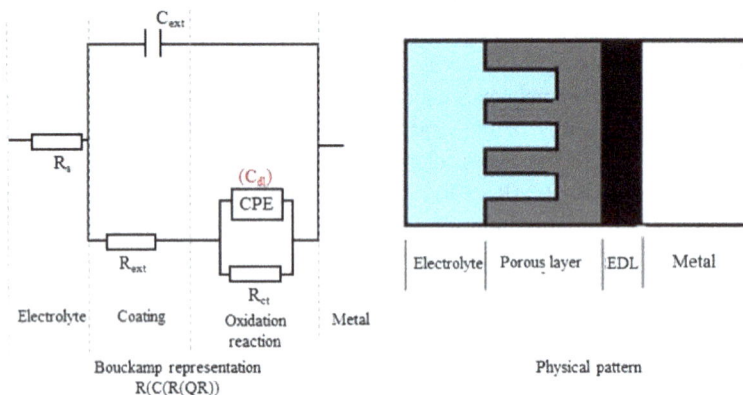

Figure 6.9 The equivalent circuit for the corrosion of a coated metal.

The outer coat of the $C45$ sample immersed in rainwater is very thin and probably non-conductive, being characterized by an extremely low electrical capacitance (3.54 nF) and high electrical resistance (1014 Ω·cm²).

The outer coat of the $C45$ sample and of the three phosphate-coated samples is even in terms of both composition and microstructure, so that the capacitance of this layer is represented by an ideal capacitor.

The electrical capacitance of the outer coat of the $C45$/APL pair is extremely low (2.54 pF). This is accounted for by the fact that, although the product layer formed in the initial moments is very thin, it is impregnated with rainwater whose conductivity is low ($\varepsilon r = 81$; $\varepsilon = \varepsilon r \cdot \varepsilon 0 = 81 \cdot 8,854.10 - 12\ F/m$). For the same reason the resistance of this layer is high. The resistance of the electrical double layer (Rct – charge transfer resistance) is also very high, hence the polarization resistance here is: $R_p = R_{ct} + R_{ext} = 4980\ \Omega.cm^2$ (Table 6. 18).

System	R_s, $[\Omega\cdot cm^2]$	C_{ext}, $[\mu F/cm^2]$	R_{ext}, $[\Omega\cdot cm^2]$	CPE		R_{ct}, $[\Omega\cdot cm^2]$	$10^3\chi^2$	εz
				Q, $[S\cdot s^n/cm^2]$	n			
C45/APL	1386	$2.54\ 10^{-3}$	1014	$1.00\ 10^{-4}$	0.65	3596	0.932	3.05
I − Zn/AMN	36.3	0.106	30.82	$6.20\ 10^{-5}$	0.51	1982	5.77	7.59
II − Zn/ Fe/AMN	40.4	2.870	11.04	$2.51\ 10^{-4}$	0.61	883	0.694	2.44
III − Mn/AMN	37.9	4.46	2.76	$5.31\ 10^{-2}$	0.73	463	0.109	1.02
OFU/APL	1969	2.06 pF	1393	$1.48\ 10^{-5}$	0.49	$2.19\ 10^4$	1.89	4.68

The capacitance of the phosphate-coated samples is of the order of microfarads, being specific to porous layers. Since the pores are filled with corrosion medium (seawater in our case), their resistance is very low. The $I − Zn$ sample immersed in seawater has the highest charge transfer resistance, as the phosphate coat deposited in the first solution provides better protection than the other two phosphate-coated samples. In this case, the frequency exponent value (n=0.51) proves that the constant phase element replacing the electrical double layer capacitance is actually a Warburg element, which also indicates the influence of diffusion on the corrosion rate.

Charge transfer resistance (R_{ct}) decreases in the following order $I − Zn > II − Zn/Fe > III − Mn$, $I − Zn$, which provides the best corrosion protection. The capacitance of the electrical double layer is represented by a non-ideal capacitor, so it was necessary to introduce the constant phase element (CPE), and the deviation from ideality is considerable.

The resistance of the outer coat and especially the resistance of the electrical double layer of the phosphate-coated and lubricated sample immersed in rainwater are very high, which is why the polarization resistance is of the order of tens of $k\Omega \cdot cm^2$ (more precisely: Rp=23.9 $k\Omega \cdot cm^2$). This is why the instantaneous corrosion rate could not be measured. Again, the constant phase element has the significance of a Warburg impedance.

Solution resistance is higher in the *OFU*/APL system than in the *C*45/APL system, although the geometry of the measuring cell is the same; this is because the resistance of the lubricant film is added to the actual resistance of the solution.

6.3.2.3 R(Q(R(QR))) equivalent circuit

The equivalent circuit shown in Fig. 6.10 describes a two-layer physical state, identical to that described by the circuit in Fig. 6.9, the only difference being that in these cases the electrical capacitance of the outer coat is described by a non-ideal capacitor, represented by the CPE1 constant phase element.

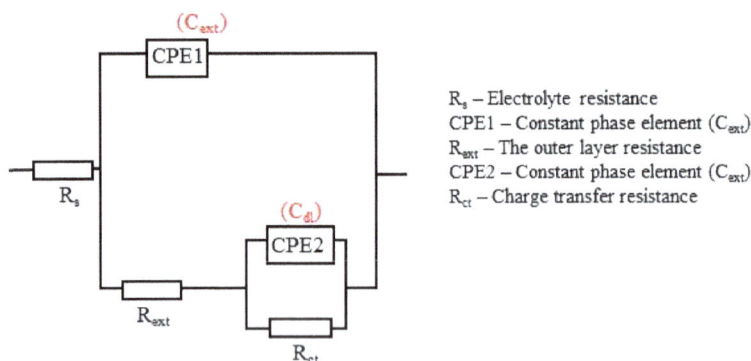

R_s – Electrolyte resistance
CPE1 – Constant phase element (C_{ext})
R_{ext} – The outer layer resistance
CPE2 – Constant phase element (C_{ext})
R_{ct} – Charge transfer resistance

Figure 6.10 The equivalent circuit R(Q(R(QR))) for all phosphate samples immersed in rainwater and fire extinguishing solution.

This circuit was suitable for processing the EIS data recorded for $I - Zn$, $II - Zn/Fe$ and $III - Mn$ after one hour of immersion in rainwater and fire extinguishing agent. The values of the parameters of the circuit elements for these systems are shown in Table 6.19.

Table 6.19 Values of the parameters of the R(Q(R(QR))) equivalent circuit elements for the C45 phosphate-coated steel samples, after one hour spent in rainwater or fire extinguishing agent.

System	R_s, $[\Omega \cdot cm^2]$	C_{ext}, $[\mu F/cm^2]$	R_{ext}, $[\Omega \cdot cm^2]$	CPE		R_{ct}, $[\Omega \cdot cm^2]$	$10^3\chi^2$	εz
				Q, $[S \cdot s^n/c \ m^2]$	n			
C45/APL	2913	$7.78 \ 10^{-10}$	0.90	4894	$9.53 \ 10^{-6}$	0.52	$3.91 \ 10^4$	0.25
I-Zn/AMN	2328	$1.63 \ 10^{-9}$	0.95	1987	$1.58 \ 10^{-4}$	0.68	1858	0.21
II-Zn/Fe/AMN	2159	$2.45 \ 10^{-9}$	0.95	1507	$1.51 \ 10^{-3}$	0.72	1357	5.52
III-Mn/AMN	4.59	$1.14 \ 10^{-3}$	0.79	115	$7.22 \ 10^{-3}$	0.64	326	0.88
OFU/APL	4.64	$3.80 \ 10^{-4}$	0.63	283	$2.03 \ 10^{-3}$	0.53	989	2.26

As with other alloys immersed in rainwater, the outer coat of phosphate-coated alloys consists of an extremely thin film saturated with rainwater so that its electrical capacitance is very low, of the order of picofarads, while resistance is very high, of the order of $k\Omega/cm^2$, this coat providing protection. Given that the frequency exponent in the expression of the CPE1 constant phase element is very close to unit, it is in fact an almost ideal capacitor.

The charge transfer resistance of the same systems is also high, so that the electrical double layer prevents the corrosion reaction from evolving. Moreover, the electrical capacitance of this coat is a non-ideal capacitor, probably due to the irregularity of the phosphate coat and to its porosity.

The polarization resistance of the three phosphate-coated samples decreases in the following order:

$$I - Zn \ (44.0 \ k\Omega \cdot cm^2) > II - Zn/Fe \ (3.85 \ k\Omega \cdot cm^2) > III - Mn \ (2.86 \ k\Omega \cdot cm^2)$$

This proves that the first solution $(I - Zn)$ phosphate-coated alloy has the best corrosion resistance in rainwater, while the third solution $(III - Mn)$ phosphate-coated alloy is the most corrosive.

The polarization resistances of the three phosphate-coated samples immersed in fire extinguishing agent are much lower and vary in the following order:

$$II - Zn/Fe \ (1.272 \ k\Omega \cdot cm^2) > I - Zn \ (0.441 \ k\Omega \cdot cm^2)$$
$$> III - Mn \ (0.133 \ k\Omega \cdot cm^2)$$

The result is in agreement with the results obtained for the instantaneous corrosion rate which is inversely proportional to the polarization resistance:

$$II - Zn/Fe(1312 \ \mu m/year) \ < \ I - Zn \ (1771 \ \mu m/year)$$
$$< \ III - Mn \ (4323 \ \mu m/year)$$

The third solution phosphate-coated alloy is the most corrosive in this corrosion medium as well, as the fire extinguishing solution is a strong corrosive agent.

The inversions that occur in the results obtained in fire extinguishing solution are most likely due to the absorption and/or adsorption of some components of the solution in the pores of the phosphate coats, whose dimensions differ from one sample to another depending on the phosphating conditions.

6.3.2.4 R(C(R(Q(RW)))) equivalent circuit

The equivalent circuit used is specific to a corrosive metal coated with a porous non-conductive film in which the corrosion process takes place under mixed, kinetic and diffusion control. Here the coating is formed by adsorption of the solution components on the surface of the freshly ground alloy. The deposited coat is compact enough to function as a barrier to charge transfer, which is now carried out by diffusion.

In this circuit, Rs is the resistance of the solution, R_{ext} and C_{ext} – the resistance and electrical capacitance of the coating layer, R_{ct} and CPE – the resistance and constant phase element replacing the electrical double layer capacitance (they control the reaction rate from a kinetic point of view), and W is the specific constant in the diffusion impedance expression, a value that controls the reaction rate by influencing the diffusion of oxidizing and reducing species to and from the metal. The parameter values of the circuit elements are shown in Table 6.20.

In the fire extinguishing solution, although the surface of the alloy is freshly ground, it is necessary to use a double layer model from the very beginning and to take into account diffusion, this is due to the fact that the solution or one of its components adheres to the surface of the metal as soon as it is immersed in the liquid and forms a very thin film. The film is porous and hence its resistance is very low and does not influence the resistance of the solution or charge transfer resistance. Moreover, the outer film is even so that the capacitance of this coat is represented by an ideal capacitor. The resistance of the electrical double layer is relatively high, much higher than the value of the polarization resistance determined from the linear polarization curve, in this case the electrochemical impedance spectroscopy method underestimates the conductive properties of the electrical double layer.

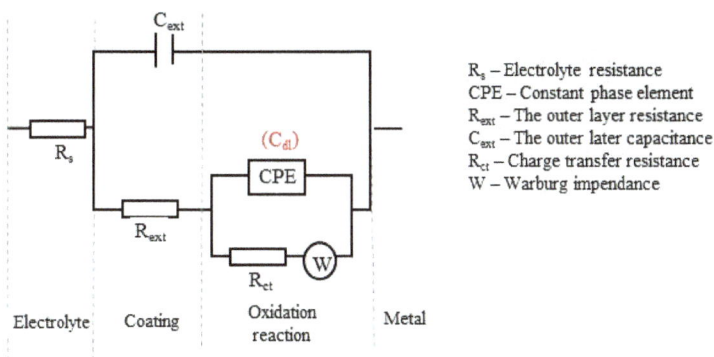

R_s – Electrolyte resistance
CPE – Constant phase element
R_{ext} – The outer layer resistance
C_{ext} – The outer later capacitance
R_{ct} – Charge transfer resistance
W – Warburg impendance

Figure 6.11 The equivalent circuit R(C(R(Q(RW)))) for a corrosive alloy under mixed kinetic and diffusion control.

Table 6.20 The values of the equivalent circuit elements for C45 immersed for one hour in fire extinguishing solution

System	R_s, $[\Omega \cdot cm^2]$	C_{ext}, $[\mu F/cm^2]$	R_{ext}, $[\Omega \cdot cm^2]$	CPE		R_{ct}, $[\Omega \cdot cm^2]$	W, [S $s^{½}/c$]	$10^3 \chi^2$	ϵz
				Q $[S \cdot s^n/cm^2]$	n				
C45/SSI	4.64	1.71	6.42	1.43 10-3	0.65	1544	0.277	0.55	2.3

6.3.2.5 R(Q(R(Q(RW)))) equivalent circuit

The circuit shown in Fig. 6.12 was used to describe the surface properties of lubricant-impregnated samples coated with phosphate in the first solution, immersed for one hour in Black Sea water and in fire extinguishing solution (*OFU*/AMN and *OFU*/SSI).

Just like the previous circuit, it describes a two-layer structure, in which the reaction rate is controlled by dual and kinetic means, and by diffusion. The difference from the previous system is that here the capacitance of the outer coat is represented by a non-ideal capacitor marked by the CPE1 constant phase element. The numerical values of the parameters of the circuit elements are shown in Table 6.21.

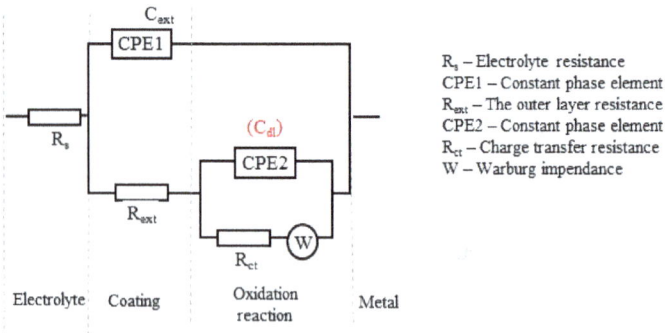

Figure 6.12   *The equivalent circuit for a corrosive alloy with kinetic and diffusion evolution of reaction rate II.*

Table 6.21   *The values of the equivalent circuit elements for OFU samples immersed in seawater and fire extinguishing solution.*

System	R_s $\Omega \cdot cm^2$	CPE 1		R_{ext} $\Omega \cdot cm^2$	CPE 2		R_{ct} $\Omega \cdot cm^2$	W $Ss^{1/2}/cm^2$	$10^3 \chi^2$	εz
		Q_1 Ss^n/cm^2	n_1		Q_2 Ss^n/cm^2	n_2				
OFU/APL	28.83	$3.16 \ 10^{-5}$	0.54	4753	$5.44 \ 10^{-5}$	0.73	3893	$5.00 \ 10^{-4}$	7.8	8.8
OFU/SSI	6.69	$6.95 \ 10^{-5}$	0.71	271	$3.63 \ 10^{-5}$	0.70	4246	$8.30 \ 10^{-4}$	8.8	9.4

The polarization resistance ($R_P = R_{ext} + R_{ct}$) of the *OFU*/AMN system ($8.65 \ k\Omega \cdot cm^2$) is two times bigger than that of the *OFU*/SSI system ($4.51 \ k\Omega \cdot cm^2$). Also, the effect of diffusion on the reduction of the corrosion rate is in favor of the *OFU*/AMN system (ZW diffusion impedance is inversely proportional to W). The sum of these (kinetic and diffusional) components triggers a sharp decline of the corrosion rate, which is incomparably lower than that of the *C*45 sample and the samples coated with phosphate in the same solutions (2→5 times in seawater and 2→6 times in SSI); the instantaneous diffusion rate is 114 μm/year in seawater and 243 μm/year in fire extinguishing solution.

6.3.2.6 R(Q(R(Q(RW)))) equivalent circuit

The equivalent circuit shown in Fig. 6.13 is an electrical circuit describing a three-layer physical state, with which the phosphate-coated and painted C45 steel samples immersed in all three corrosion media were processed very well.

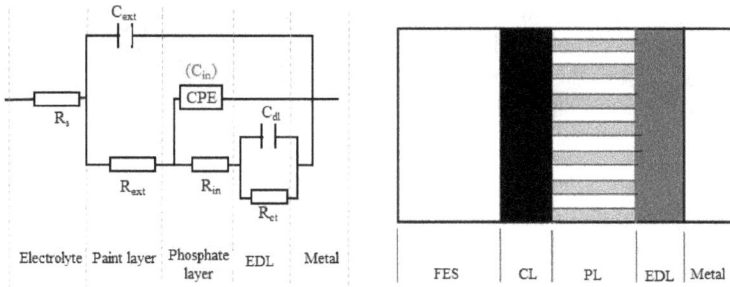

Figure 6.13   The equivalent circuit R(C(R(Q(R(CR))))) for EIS data fitting for the OFV samples immersed in APL, AMN and SSI.

Table 6.22   The values of the equivalent circuit for OFV samples immersed in APL, AMN and SSI.

System	R_s $\Omega \cdot cm^2$	C_{ext} $\mu F/cm^2$	R_{ext} $\Omega \cdot cm^2$	CPE Q $S\ s^n/cm^2$	n	R_{in} $\Omega \cdot cm^2$	C_{dl} $\mu F/cm^2$	R_{ct} $\Omega \cdot cm^2$	$10^3 \chi^2$	εz
OFV/APL	1499	0.73 pF	4709	$3.83\ 10^{-4}$	0.59	521	161.7	5208	0.86	2.95
OFV/AMN	107.1	0.175	19.12	$6.51\ 10^{-4}$	0.63	16.5	20.4	1179	0.58	2.41
OFV/SSI	16.8	52.8	119	$1.80\ 10^{-4}$	0.76	576	175	1604	0.47	2.17

The outer coat is the paint film, it has very low electrical conductivity and, although it is very thin, it has high resistance and very high electrical capacitance in rainwater, yet very low resistance in seawater, in which it seems that this coat deteriorates. This conclusion is also supported by the results of linear polarization measurements, where it was found that the corrosion rate of this sample is higher than those of the other samples immersed in seawater. Initially, this coat is a compact coat (SC).

The inner coat is the phosphate coat, which is porous (SP). The resistance of this coat is lower than in the $I - Zn$ sample, because due to the paint coat the pores are no longer clogged with fire extinguishing solution components or with reaction products (the reaction rate is very low and the initial number of products is very small). The electrical capacitance of this coat is represented by non-ideal capacitors as a consequence of the marked roughness of the surface of this coat.

Charge transfer resistance (R_{ct}) is very high in rainwater and, compared to the resistance of the other coats, polarization resistance is very high ($R_P = R_{ext} + R_{in} + R_{ct} = 10.4$ kΩ·cm^2). This is the reason why the linear polarization method could not record any data about OFV corrosion in rainwater. When the sample is immersed in seawater, R_{ct} has the lowest value and hence the polarization resistance is also very low (R_P=1.22 kΩ·cm^2) and the paint film does not provide satisfactory protection in seawater (Table 6.22).

The degradation of the coated metals was assessed using the Electrochemical Impedance Spectroscopy method. The experimental data collected for the six types of samples exposed to three corrosion media were processed with three equivalent cell types:

(i) an equivalent circuit for an alloy whose corrosion is controlled only kinetically (the corrosion rate is controlled only by the penetration of the electrical double layer by active species);

(ii) an equivalent circuit for metals (alloys) coated with relatively porous layers, in which the corrosion rate is controlled only kinetically or dually; kinetically and by diffusion;

(iii) an equivalent circuit that describes a three-layer physical state: EDL, porous layer, and compact layer.

The analysis of the experimental data based on the curves adjusted using the equivalent circuits enabled us to conclude that the surface structure and behavior of the $C45$ steel have different aspects in the three corrosion media employed.

In rainwater, the EIS analysis of the $C45$ sample indicates a two-layer structure in which corrosion is controlled only kinetically. Measurements revealed an extremely thin layer of formed products, which is saturated with rainwater and whose electrical conductivity is very low. Under these circumstances, the electrical capacitance of the outer coat is very low (of the order of picofarads), while its resistance is very high. On the other hand, the electrical double layer has high resistance, so its polarization resistance is also high (Rp =4980 Ω·cm2) and its corrosion rate is very low.

In Black Sea water, due to the aggressiveness of chlorine ions attacking the surface of the $C45$ sample (mainly the iron in the steel) no continuous layer is initially formed but granules of products (insoluble oxides or oxy-hydroxides), part of which migrate into the solution, and the others are adsorbed on the surface. Therefore, the Nyquist diagram shows a negative loop within the low frequency range, hence a solenoid and resistance R_L need to be introduced in the equivalent circuit. Under these circumstances, the polarization resistance is Rp=1503 Ω·cm^2, i.e. 3.3 times lower than in rainwater, the corrosion rate being higher.

When immersing the $C45$ sample in fire extinguishing agent, the equivalent circuit used is specific of a corrodible metal coated with a porous non-conductive film in which the corrosion process takes place under mixed kinetic and diffusion control. This may be accounted for by supposing that SSI or some of its components adhere to the surface of the metal as soon as it is immersed into the liquid thus forming a very thin film. The layer thus formed is porous and hence its resistance is very low and does not influence the solution resistance or the charge transfer resistance. Moreover, the outer film is even so that its capacitance is represented by an ideal capacitor. The resistance of the electrical double layer is relatively high, much higher than the polarization resistance determined from the linear polarization curve. However, the estimated corrosion rate resulting from the EIS data is of the same order of magnitude as in the case of seawater.

The three phosphate-coated samples immersed in a given corrosion medium may be processed using the same equivalent circuit, the difference between them being the values of the circuit parameters that express different physical states.

In rainwater, the surface structure of the three phosphate-coated samples corresponds to a two-layer structure, and the corrosion process is controlled only kinetically. The phosphate coat on each sample is porous, the pores being initially filled with rainwater, so that the resistance of the outer coat is also high. The capacitance of the outer coat is low (of the order of nanofarads) and the circuit element approaches an ideal capacitor. The electrical double layer is characterized by very high resistance but is uneven. Under these circumstances, the polarization resistance is high (44 $k\Omega \cdot cm^2$ - for $I - Zn$, 3.85 $k\Omega \cdot cm2$- for $II - Zn/Fe$ and 2.86 $k\Omega \cdot cm^2$ – for $III - Mn$). These data prove that the sample coated with phosphate in the first solution (containing $Zn_3(PO_4)_2$) has the best corrosion resistance in rainwater – over 10 times better than the other two samples, while the sample coated with phosphate in the third solution (containing $Mn_3(PO_4)_2$+Fe+Ni) is the most corrodible.

In Black Sea water, the surface structure of phosphate-coated samples is also a two-layer structure, the equivalent circuit being similar to that used for rainwater testing, the only difference being that in this case the electrical capacitance of the outer coat is represented by an ideal capacitor instead of the phase element, a constant whose value is 103 ÷ 104 times higher. The resistances of the outer coat and the resistances of the electrical double layer are much lower than in rainwater and the corrosion rates are higher. However, as in the previous case, the sample coated with phosphate in the first solution provides the best protection, while the sample coated with phosphate in the third solution is the least resistant to corrosion.

In fire extinguishing solution, the data processing for the phosphate-coated samples was performed with the same circuit as in the case of rainwater. In this case, the capacitance of the outer coat is represented by a capacitor very far from the ideal case, probably due to the absorption in the pores of some SSI components. The resistance of the outer coat is low because the pores of the phosphate coats are filled with a solution which has very high electrical conductivity. The electrical double layer is uneven and the polarization resistance exhibits an inversion of values compared to other corrosion media. Here, the $II - Zn/Fe$ sample appears as the most resistant to corrosion, while the $III - Mn$ sample as the most corrodible.

The experimental results concerning the rainwater behavior of samples coated with phosphate in the first solution and impregnated in lubricant were interpreted with the same equivalent circuit used to process the data for the $C45/\text{APL}$ system, a circuit that describes a two-layer physical state with kinetic corrosion rate control. The difference between this system and the $C45/\text{APL}$ system consists of the considerable discrepancies between the outer coat resistances and the charge transfer resistance, so that in this case the polarization resistance is 23.29 $k\Omega\cdot cm^2$ and for $C45/\text{APL}$ it is 4.98 $k\Omega\cdot cm^2$ (i.e. 4.7 times higher). This makes the corrosion rate very low, virtually negligible. The result is in agreement with the result of the linear polarization measurements, where the curve could not be recorded due to the very low currents - of the order of picoamperes.

In AMN and in SSI, the EIS data for the OFU sample were processed using the same equivalent circuit. It describes a two-layer physical system with mixed kinetic and diffusion corrosion rate control. In these two media, both the capacitance of the electrical double layer and the capacitance of the outer coat are represented by non-ideal capacitors, this indicating both a rough structure (EDL) and a porous and uneven structure from a chemical point of view (for the outer coat). Although the outer coat consists of the deposited phosphate film and of the film of adsorbed products from the solution, it appears as a unitary layer, despite its composition not being even (the frequency exponent in the expression of constant phase elements is far from the unit). Capacitances (expressed by the CPE1 and CPE2 constant phase elements), resistances (R_{ext} and R_{ct}) and Warburg constant (W) recorded for OFU in the two corrosion media are of the same order of magnitude. Considering the very low values of the W constant, diffusion impedance (ZW) is very high. The kinetic ($R_{ct} + R_{ext}$) and diffusion (ZW) components taken together provide very high polarization resistance and trigger a sharp decline of the corrosion rate, which is incomparably lower than for the $C45$ sample and for the phosphate-coated samples (2→5 times in seawater and 2→6 times in fire extinguishing agent); the instantaneous diffusion rate is 114 µm/year in seawater and 243 µm/year in SSI.

The painted phosphate-coated sample was analyzed in all three corrosion media with the same equivalent electrical circuit, which describes a three-layer physical state with kinetic corrosion rate control. The outer coat is compact and consists of the paint film, the inner coat is porous and consists of the zinc phosphate coat, while the third coat is the electrical double layer.

The charge transfer resistance (R_{ct}) and polarization resistance of the OFV sample are very high in rainwater compared to the other analyzed samples. This difference is to blame for the inability to measure the corrosion of the *OFV* sample in rainwater, using the linear polarization method.

When the *OFV* sample is immersed in seawater, Rct has the lowest value and hence its polarization resistance is also very low (RP=1.22 kΩ·cm^2); consequently, the coat of paint is severely damaged by this medium and does not provide satisfactory protection.

The painted sample is better protected in fire extinguishing solution than in seawater, as the resistances involved (R_{ext}, R_{in} and R_{ct}) are higher. This may be a consequence of the adsorption of the solution on the surface.

The aggressiveness of corrosion media is determined both by their composition (aggressive ions, neutral molecule inhibitors, etc.) and by their electrical conductivity. The value of the resistance of the solution (R_s – the resistance of the corrosion medium between the sample surface and the reference electrode plus the resistance of the outer coats) may be considered a relative measure of the conductivity of the corrosion media used in this research. The values of this resistance are comparable, since the same cell geometry was used throughout the measurements. The mean Rs values determined in the *C*45 sample and in the phosphate-coated samples were analyzed for this purpose.

The fire extinguishing solution is the best conductor of electricity, its mean solution resistance value being $R_s = 4.71\pm0.14$ Ω·cm^2, the conductivity of seawater is eight times lower, $R_s = 38.55\pm1.45$ Ω·cm^2.

The Rs values of the *OFU* and *OFV* samples deviate greatly from the mean value due to the very high resistances of the lubricant and of the paint which are added to the actual value of the solution. Thus, for the *OFU* sample, Rs is 6.69 Ω·cm^2- in SSI and 28.83 Ω·cm^2 – in AMN, while for the painted sample Rs is 16.8 Ω·cm^2 – in SSI and 107.1 Ω·cm^2 in AMN.

In rainwater, the solution resistance values are very scattered, ranging from 1386 Ω·cm^2 to 2913 Ω·cm^2 (mean value for all systems: 1726$\pm$545 Ω·cm^2). These differences are due to the fact that rainwater has extremely low conductivity (close to that of distilled water) and small variations in the ER/sample distance considerably affect resistance.

References

[1]   N. Etteyeb, M. Sanchez, L. Dhouibi, C. Alonso, C. Andrade, E. Triki, Corrosion protection of steel reinforcement by a pretreatment in phosphate solutions: Assessment of passivity by electrochemical techniques. Corrosion Engineering Science and Technology, 41 (2006) 336 341. https://doi.org/10.1179/174327806X120775

[2]   D.P. Burduhos-Nergiş, C. Nejneru, R. Cimpoeşu, A.M. Cazac, C. Baciu, D.C. Darabont, C. Bejinariu, Analysis of chemically deposited phosphate layer on the carabiners steel surface used at personal protective equipments. Quality - Access to Success, 20 (2019) 77 82.

[3]   D.P. Burduhos-Nergis, P. Vizureanu, A.V. Sandu, C. Bejinariu, Phosphate Surface Treatment for Improving the Corrosion Resistance of the C45 Carbon Steel Used in Carabiners Manufacturing, Materials 13(15) (2020) 3410. https://doi.org/10.3390/ma13153410

[4]   C. Bejinariu, I. Sandu, V. Vasilache, I.G. Sandu, M.G. Bejinariu, A.V. Sandu, M. Sohaciu, V. Vasilache, Process for the micro-crystalline phosphate-coating of iron-based metal pieces, Patent RO125457-B1/2014.

[5]   C.R. Tomachuk, C.I. Elsner, A.R. di Sarli, Electrochemical characterization of chromate free conversion coatings on electrogalvanized steel, Materials Research 17 (2014) 61–68. https://doi.org/10.1590/S1516-14392013005000179

[6]   C. Nejneru, M.C. Perju, D.D. Burduhos Nergis, A.V. Sandu, C. Bejinariu, Galvanic Corrosion Behaviour of Phosphate Nodular Cast Iron in Different Types of Residual Waters and Couplings, REVISTA DE CHIMIE, 70(10) (2019) 3597-3602. https://doi.org/10.37358/RC.19.10.7604

[7]   P. Lazar, C. Bejinariu, A.V. Sandu, A.M. Cazac, O. Corbu, M.C. Perju, I.G. Sandu, Corrosion Evaluation of Some Phosphated Thin Layers on Reinforcing Steel, IOP Conference Series: Materials Science and Engineering, 209(1) (2017) 012025. https://doi.org/10.1088/1757-899X/209/1/012025

[8]   C.A. da Cunha, N.B. de Lima, J.R. Martinelli, A.H. de A. Bressiani, A.G.F. Padial, L.V. Ramanathan, Microstructure and mechanical properties of thermal sprayed nanostructured Cr_3C_2-$Ni_{20}Cr$ coatings. Materials Research, 11 (2008) 137–143. https://doi.org/10.1590/S1516-14392008000200005

[9]   A. Popoola, O. Olorunniwo, O. Ige, Corrosion Resistance Through the Application of Anti- Corrosion Coatings. In Developments in Corrosion Protection; InTech,

2014. https://doi.org/10.5772/57420

[10]   K.P. Balan, Chapter Nine - Corrosion. Metallurgical Failure Analysis, 155–178, 2018. https://doi.org/10.1016/B978-0-12-814336-0.00009-3

[11]   D. Dwivedi, K. Lepková, T. Becker, Carbon steel corrosion: a review of key surface properties and characterization methods. RSC Advances 7 (2017) 4580–4610. https://doi.org/10.1039/C6RA25094G

[12]   V. de Freitas Cunha Lins, G.F. de Andrade Reis, C.R. de Araujo, T. Matencio, Electrochemical impedance spectroscopy and linear polarization applied to evaluation of porosity of phosphate conversion coatings on electrogalvanized steels. Applied Surface Science, 253 (2006) 2875–2884. https://doi.org/10.1016/j.apsusc.2006.06.030

[13]   A.V. Sandu, A. Ciomaga, G. Nemtoi, C. Bejinariu, I. Sandu, Study on the chemical deposition on steeel of zinc phosphate with other metallic cations and hexamethilen tetramine. II. Evaluation of corrosion resistance, Journal of Optoelectronics and Advanced Materials, 14(7-8) (2012) 704-708.

[14]   J.M. Blengino, M. Keddam, J.P. Labbe, L. Robbiola, Physico -chemical characterization of corrosion layers formed on iron in a sodium carbonate-bicarbonate containing environment. Corrosion Science, 37(4) (1995) 621–643. https://doi.org/10.1016/0010-938X(94)00160-8

[15]   L. Dhouibi, E. Triki, M. Salta, P. Rodrigues, A. Raharinaivo, Studies on corrosion inhibition of steel reinforcement by phosphate and nitrite. Materials and Structures, 36 (2003) 530–540. https://doi.org/10.1007/BF02480830

[16]   D.P. Burduhos Nergis, C. Nejneru, D.D. Burduhos Nergis, C. Savin, A.V. Sandu, S.L. Toma, C. Bejinariu, The galvanic corrosion behavior of phosphated carbon steel used at carabiners manufacturing. Revista de Chimie, 70 (2019) 215–219. https://doi.org/10.37358/RC.19.1.6885

[17]   A.V. Sandu, C. Bejinariu, G. Nemtoi, I.G. Sandu, P. Vizureanu, I. Ionita, C. Baciu, New anticorrosion layers obtained by chemical phosphatation, REVISTA DE CHIMIE, 64(8) (2013) 825-827.

[18]   P.-E. Nica, M. Agop, S. Gurlui, C. Bejinariu, C. Focsa, Characterization of Aluminum Laser Produced Plasma by Target Current Measurements. Jpn. J. Appl. Phys. 51 (2012) 106102. https://doi.org/10.1143/JJAP.51.106102

[19]   C. Bejinariu, D.P. Burduhos-Nergis, N. Cimpoesu, M.A. Bernevig-Sava, S.L. Toma, D.C. Darabont, C. Baciu, Study on the anticorrosive phosphated steel

carabiners used at personal protective equipment, Quality-Access to Success 20(1) (2019) 71-76.

[20] B.V. Jegdić, J.B. Bajat, J.P. Popić, S.I. Stevanović, V.B. Mišković-Stanković, The EIS investigation of powder polyester coatings on phosphated low carbon steel: The effect of NaNO2 in the phosphating bath. Corrosion Science, 53 (2011) 2872–2880. https://doi.org/10.1016/j.corsci.2011.05.019

[21] M. Manna, Characterisation of phosphate coatings obtained using nitric acid free phosphate solution on three steel substrates: An option to simulate TMT rebars surfaces. Surface and Coatings Technology, 203 (2009) 1913–1918. https://doi.org/10.1016/j.surfcoat.2009.01.024

[22] M. Chen, C.Y. Du, G.P. Yin, P.F. Shi, T.S. Zhao, Numerical analysis of the electrochemical impedance spectra of the cathode of direct methanol fuel cells. Internationa Journal of hydrogen energy, 34 (2009) 1522–1530. https://doi.org/10.1016/j.ijhydene.2008.11.072

[23] V.H. Radosevic, K. Kvastek, D. Hodko, V. Pravdi´c, Impedance of anodically passivated FeB over potentials from passive state to oxygen evolution. Electrochimica Acta, 39(1) (1994) 119–130. https://doi.org/10.1016/0013-4686(94)85018-6

[24] M. Gaberscek, S. Pejovnik, Impedance spectroscopy as a technique for studying the spontaneous passivation of metals in electrolytes. Electrochimica Acta, 41(7-8) (1996) 1137-1142. https://doi.org/10.1016/0013-4686(95)00464-5

[25] E.B. Castro, J.R. Vilche, Investigation of passive layers on iron and iron-chromium alloys by electrochemical impedance spectroscopy. Electrochimica Acta, 38 (11) (1993) 1567– 1572. https://doi.org/10.1016/0013-4686(93)80291-7

[26] D.P. Burduhos-Nergis, P. Vizureanu, A.V. Sandu, C. Bejinariu, Evaluation of the Corrosion Resistance of Phosphate Coatings Deposited on the Surface of the Carbon Steel Used for Carabiners Manufacturing, Applied Sciences 10(8) (2020) 2753. https://doi.org/10.3390/app10082753

[27] E.S. Bacaita, C. Bejinariu, B. Zoltan, C. Peptu, G. Andrei, M. Popa, D. Magop M. Agop, Nonlinearities in Drug Release Process from Polymeric Microparticles: Long-Time-Scale Behaviour, J. Appl. Math. 653720, 2012. https://doi.org/10.1155/2012/653720

[28] B.-Y. Chang, S.-M. Park, Electrochemical Impedance Spectroscopy, Annual Review of Analytical Chemistry, 3 (2010) 207-229.

https://doi.org/10.1146/annurev.anchem.012809.102211

[29]   C. Bejinariu, P. Lazăr, A.V. Sandu, A.M. Cazac, I.G. Sandu, O. Corbu, Enhancing
        properties of reinforcing steel by chemical phosphatation, Applied Mechanics and
        Materials, 754-755 (2015) 310-314.
        https://doi.org/10.4028/www.scientific.net/AMM.754-755.310

[30]   A.M. Nagiub, Evaluation of Corrosion Behavior of Copper in Chloride Media
        Using Electrochemical Impedance Spectroscopy (EIS). Port. Electrochim. Acta, 23
        (2005) 301-314. https://doi.org/10.4152/pea.200502301

[31]   E. Cano, D. Lafuente, D.M. Bastidas, Use of EIS for the evaluation of the
        protective propertie of coatings for metallic cultural heritage: a review. J Solid
        State Electrochem, 14 (2010) 381-391. https://doi.org/10.1007/s10008-009-0902-6

[32]   F. Mansfeld, An Introduction to Electrochemical Impedance Spectroscopy.
        Technical Report No. 26, Solartron Limited, Los Angeles, United States of
        America, 1999.

[33]   In.A. Lasia, R.E. White, B.E. Conway, J.O.M. Bockris, Modern Aspects of
        Electrochemistry 32, Kluwer Academic/Plenum Publishers, New York, United
        States of America, 1999.

[34]   G.J. Brug, A.L.G. Van Den Eeden, M. Sluyters-Rehbach, J.H. Sluyters, The
        analysis of electrode impedances complicated by the presence of a constant phase
        element. J. Electroanal. Chem. 176 (1984) 275. https://doi.org/10.1016/S0022-
        0728(84)80324-1

[35]   J. Ross Macdonald, Impedance Spectroscopy: Emphasizing Solid Materials and
        Analysis, John Wiley & Sons, New York, United States of America, 1987.

[36]   M. Itagaki, A. Taya, K. Watanabe, K. Noda, Deviations of Capacitive and
        Inductive Loops in the Electrochemical Impedance of a Dissolving Iron Electrode.
        Analytical Science, 18 (2002) 641-644. https://doi.org/10.2116/analsci.18.641

[37]   P. Zoltowski, On the electrical capacitance of interfaces exhibiting constant phase
        element behavior. J. Electroanal. Chem., 443 (1998) 149-154.
        https://doi.org/10.1016/S0022-0728(97)00490-7

[38]   V. Horvat-Radošević, K. Kvastek, D. Hodko, V. Pravdić, Impedance of anodically
        passivated Fe80B20 over potentials from passive state to oxygen evolution.
        Electrochimica Acta, 39 (1994) 119–130. https://doi.org/10.1016/0013-
        4686(94)85018-6

[39]   D.P. Burduhos-Nergis, C. Bejinariu, S.L. Toma, A.C. Tugui, E.R. Baciu, Carbon

steel carabiners improvements for use in potentially explosive atmospheres. MATEC Web of Conferences, 305, 00015, 2020. https://doi.org/10.1051/matecconf/202030500015

[40] D.P. Burduhos-Nergis, A.M. Cazac, A. Corabieru, E. Matcovschi, C. Bejinariu, Characterization of Zinc and Manganese Phosphate Layers Deposited on the Carbon Steel Surface, Euroinvent ICIR IOP Conference Series: Materials Science and Engineering 877 (2020) 012012. https://doi.org/10.1088/1757-899X/877/1/012012

[41] D.P. Burduhos Nergis, N. Cimpoesu, P. Vizureanu, C. Baciu, C. Bejinariu, Tribological characterization of phosphate conversion coating and rubber paint coating deposited on carbon steel carabiners surfaces, Materials today: proceedings 19 (2019) 969-978. https://doi.org/10.1016/j.matpr.2019.08.009

[42] D.P. Burduhos-Nergis, A.V. Sandu, D.D. Burduhos-Nergis, D.C. Darabont, R.-I. Comaneci, C. Bejinariu, Shock Resistance Improvement of Carbon Steel Carabiners Used at PPE, MATEC Web Conf. 290 (2019) 12004. https://doi.org/10.1051/matecconf/201929012004

[43] R.A. Cottis, S. Turgoose, R. Neuman, Corrosion testing made easy Impedance and Noise, Chapter 3. NACE International, Houston, Texas, United States of America, 1999.

[44] S. Nakayama, Mechanistic stud by electrochemica impedance spectroscopy on reduction of copper oxides in neutral solutions. SEI Technical Review, 68 (2009) 62-68.

[45] C. Cao, J. Zhang, Introduction of electrochemical impedance spectroscopy. Science Press, ISBN: 7-03-009854-4, Beijing, China, 2002.

CHAPTER 7

Design and Applications

Phosphate-coating is a chemical process in which coatings are generated on the surface of a pure metal (usually steel). This consists essentially in the formation on the metal surface of a protective film of insoluble phosphates. The coatings generated by phosphating are electrically non-conductive and therefore decrease the value of the corrosion current. The coatings are not soluble in water and organic solvents and properly adhere to the base metal [1-3].

7.1 Classification

Depending on the solution and the formation conditions, phosphate coatings are of two types: crystalline and amorphous. The widest scope is covered by crystalline coatings, which have ferrous materials as their substrate (carbon steel, low alloy steel or cast iron) [4,5]. Phosphating of non-ferrous metals, stainless steels as well as amorphous phosphating are applied in special cases, the working technologies being specific to each case [6].

Crystalline phosphate coating

This type of phosphate coating is achieved starting from the slightly acidic aqueous solutions of monometallic phosphates of bivalent metals: iron, manganese and zinc, the contact with the metal surface being deposited by immersion or spraying (depending on the shape or size of the part), at temperatures ranging between 25 and 98°C [7].

The process is slow in simple phosphate solutions, and the phosphate-coating operation takes a long time. Therefore, accelerating substances are added to the basic components in most phosphate-coating solutions. Various metals (copper, nickel) or oxidizing agents, such as: nitrates, nitrites, chlorates or hydrogen peroxide may be added in controlled quantities for this purpose. In this case, it should be borne in mind that the phosphate coats obtained from solutions in which accelerators were added have low properties compared to accelerator-free solutions. Therefore, they are generally used only if the achieved layer is to be coated with paint or varnish [8-10].

Organic substances containing ethylenediamine ($C_2H_4(NH_2)_2$) are also added to phosphate-coating solutions in addition to accelerators, which are designed to reduce the amount of sludge that forms in a phosphating bath. Also, when phosphate coating using zinc solutions, alkaline earth phosphates or polyphosphates are added to reduce the crystals [10]. When poured into the solution, they increase the corrosion resistance of the coat and its resistance to mechanical shocks.

Nickel added to manganese phosphate baths increases the hardness of the deposited coat and its ability to retain lubricants [11].

Depending on the surface structure of the base metal, phosphate crystals have a different orientation, and a distinct phosphate coat morphology. For instance, a 2-4 nm thick microcrystalline film is formed on polished steel. In order to obtain a macrocrystalline structure, the time of contact of the solution with the part must be longer [12-14].

Amorphous phosphate coating

This particular type of phosphating is performed starting from solutions containing salts of different nature: alkaline ammonium phosphates or sometimes organic phosphates, without initially incorporating free phosphoric acid [15].

The hydrolysis phenomenon occurs in solutions at a higher pH than in the case of acid phosphating. In this case, a substitution reaction takes place instead of the precipitation reaction, where the cations present in the solution (sodium or ammonium) do not occur in the coating, which essentially consists of ferric phosphate $FePO_4 - 2H_2O$, metal oxide containing iron oxide Fe_2O_3. Thus, the base metal alone contributes to the formation of the coat, without any intervention from other metal elements. The film thus obtained, sensitive to mechanical stress (accidental deformation or shock), is a good base for painting (usually sheet metal). When surfactants are added to these solutions, metal degreasing may be done simultaneously with the phosphating process [16,17].

7.2 Properties of the phosphate layer

Coat properties are particularly important. Thus, they are sometimes used instead of metal coatings. The most important properties of phosphate films are:

   - high electrical resistance (phosphate films are good electrical insulators even in thin 1-15 micron coats) [9];

   - phosphate films withstand temperatures of up to about 500°C;

   - phosphate films increase the adhesion properties of oil and paint coats [18];

   - high corrosion resistance [19-24];

   - the porosity of the phosphate film is high (about 0.5%), which makes these parts suitable for impregnation in lubricant or painting immediately after phosphate coating [25];

- phosphate coats do not change the properties of the base metal, such as: hardness, elasticity, magnetic properties, etc.;

- the weight of the phosphate-coated part is not altered by phosphate-coating [26];

- the phosphate film appears matte, gray or black, as the color of the phosphate film varies depending on the composition of the phosphating solution used, the films with a higher iron content are almost black, while zinc phosphate films are lighter than those containing iron or manganese phosphate [7.8];

- the flexibility and elasticity of phosphate films is low;

- friction resistance is low [25];

7.3 Applications of phosphate coating

Phosphate-coating used as corrosion protection

Phosphate coating provides efficient physical protection to metals prone to corrosion. Thus, due to its insulating nature, phosphate coating prevents its occurrence and spread. In the automotive industry, phosphating is used to provide corrosion protection to mass-produced machine parts. Moreover, the crystalline nature of phosphate coats enhances paint adhesion, which is significantly higher than when the paint is applied directly on the metal [20-24].

Phosphate coating before cold working of metals

Steel extrusion has been widely used in industry only after solutions were found to all the lubrication problems involved. The main problem with extrusion is the separation of the surfaces in order to avoid a metal contact between the part and the mold, as well as lubrication designed to reduce friction. Due to the extremely high pressures that occur during deformation, the lubricant tends to be removed. By coating the parts with a separating film, which is also a good lubricant carrier, the requirements for low-cost extrusion have been achieved. In this application, the phosphate coats combined with suitable lubricants act as intermediate layers that reduce friction between the tool and the formed part [27].

Phosphate coating as a surface treatment for paint application

Phosphate coats prevent rusting of the layer under the paint and improve the adhesion of the paint to the metal surface, which results in better corrosion resistance of the painted phosphate-coated surface [19, 25].

In terms of corrosion protection, the application of a thick phosphate coat is the most efficient. However, these coats cause individual crystal losses and hence cracks in the paint films. Therefore, the phosphate coat should be as thin as possible.

Phosphate coating of moving parts before functional tests

Phosphate coating is a method of reducing the wear of various moving elements and parts. Phosphate coatings act as lubricants; moreover, their ability to retain oils further improves this property. Manganese phosphates, supplemented with appropriate lubricants, are most commonly used to improve wear resistance. Manganese phosphates widely used in the automotive industry are best for improving slip and reducing the associated wear of two steel surfaces that rub against each other. Phosphate coatings do not have intrinsic lubricating properties, but they can absorb or retain a considerable amount of lubricant within their pores. This combination favors an easier transition to higher surface pressures by forming a non-metallic barrier that separates the two metal surfaces and reduces the risk of cracking [28].

Phosphate coating of parts in military industry

Parkerizing is a process usually considered improved by zinc or manganese phosphate coating, although some use the term parkerizing as a generic term for the application of phosphate coatings that include the iron phosphating process. Bondering and phosphating are other terms associated with the parkerizing process.

Parkerizing is commonly used on firearms, which is a previously developed chemical conversion coating. It is also widely used in automobiles to protect unfinished metal parts against corrosion. The Parkerizing process cannot be used on non-ferrous metals such as aluminum, brass or copper. Similarly, it cannot be applied to steels that contain a large amount of nickel or to stainless steel [29].

Phosphate coating is a chemical process by which a metal, cast iron or steel surface, previously cleaned of all non-metal, is immersed (or sprayed on) in a solution of primary zinc or manganese phosphate, which may also contain other oxidizing compounds, activators or accelerators at a suitable temperature. They are coated with a layer of insoluble microcrystalline phosphate consisting of tertiary zinc and iron phosphate or of zinc or manganese phosphates [1-3].

Depending on the medium and the formation conditions, phosphate coatings are of two types: crystalline and amorphous.

In addition to the properties listed above, phosphate coatings are electrically insulating, which allows their use in the electrical engineering industry.

The process is simple and cost effective because it does not involve expensive raw materials, complicated equipment and highly qualified personnel. All these properties of phosphate films have led to the rapid development of the process and the widening of its scope.

References

[1]   A.V. Sandu, C. Bejinariu, I.G. Sandu, M.M.A.B. Abdullah, Modern Technologies of Thin Films Deposition. Chemical Phosphatation, Material Research Forum, USA (ISBN 978-1-945291-90-6) (2018) 149.

[2]   C. Bejinariu, D.P. Burduhos-Nergis, N. Cimpoesu, M.A. Bernevig-Sava, S.L. Toma, D.C. Darabont, C. Baciu, Study on the anticorrosive phosphated steel carabiners used at personal protective equipment, Quality-Access to Success 20(1) (2019) 71-76.

[3]   Y.L. Cheng, H.I. Wu, Z.H. Chen, H.M. Wang, L.L. Ling, Phosphating process of AZ31 magnesium alloy and corrosion resistance of coatings, Transactions of Nonferrous Metals Society of China 16(5) (2006) 1086-1091. https://doi.org/10.1016/S1003-6326(06)60382-8

[4]   D.P. Burduhos-Nergis, A.M. Cazac, A. Corabieru, E. Matcovschi, C. Bejinariu, Characterization of Zinc and Manganese Phosphate Layers Deposited on the Carbon Steel Surface, Euroinvent ICIR IOP Conference Series: Materials Science and Engineering 877 (2020) 012012. https://doi.org/10.1088/1757-899X/877/1/012012

[5]   X.J. Cui, C.-H. Liu, R.S. Yang, Q.S. Fu, X.Z. Lin, M. Gong, Duplex-layered manganese phosphate conversion coating on AZ31 Mg alloy and its initial formation mechanism, Corrosion Science 76 (2013) 474-485. https://doi.org/10.1016/j.corsci.2013.07.024

[6]   T.S.N. Sankara Narayanan, Surface pretretament by phosphate conversion coatings - A review., Advanced Materials Science 9 (2005) 130-177.

[7]   A.V. Sandu, A. Ciomaga, G. Nemtoi, C. Bejinariu, I. Sandu, SEM-EDX and microFTIR studies on evaluation of protection capacity of some thin phosphate layers, Microscopy Research and Technique, 75(12) (2012) 1711-1716. https://doi.org/10.1002/jemt.22120

[8]   A.V. Sandu, C. Coddet, C. Bejinariu, Study on the chemical deposition on steeel of zinc phosphate with other metallic cations and hexamethilen tetramine. I. Preparation and structural and chemical characterization, Journal of Optoelectronics and Advanced Materials, 14(7-8) (2012) 699 - 703.

[9]   D.P. Burduhos-Nergis, C. Bejinariu, S.L. Toma, A.C. Tugui, E.R. Baciu, Carbon
        steel carabiners improvements for use in potentially explosive atmospheres.
        MATEC Web of Conferences, 305, 00015, 2020.
        https://doi.org/10.1051/matecconf/202030500015

[10]   A.C. Bastos, M.G.S. Ferreira, A.M. Simões, Comparative electrochemical studies of
        zinc chromate and zinc phosphate as corrosion inhibitors for zinc., Progress in
        Organic Coatings 52(4) (2005) 339-350.
        https://doi.org/10.1016/j.porgcoat.2004.09.009

[11]   E.P. Banczek, P.R.P. Rodrigues, I. Costa, The effects of niobium and nickel on the
        corrosion resistance of the zinc phosphate layers, Surface and Coatings
        Technology, 202(10) (2008) 2008-2014.
        https://doi.org/10.1016/j.surfcoat.2007.08.039

[12]   A.V. Sandu, A. Ciomaga, G. Nemtoi, C. Bejinariu, I. Sandu, Study on the
        chemical deposition on steeel of zinc phosphate with other metallic cations and
        hexamethilen tetramine. II. Evaluation of corrosion resistance, Journal of
        Optoelectronics and Advanced Materials, 14(7-8) (2012) 704-708.

[13]   A.V. Sandu, C. Coddet, C. Bejinariu, A Comparative Study on Surface Structure
        of Thin Zinc Phosphates Layers Obtained Using Different Deposition Procedures
        on Steel, REVISTA DE CHIMIE, 63(4) (2012) 401-406.

[14]   C. Bejinariu, I. Sandu, V. Vasilache, I.G. Sandu, M.G. Bejinariu, A.V. Sandu,
        M. Sohaciu, V. Vasilache, Process for the micro-crystalline phosphate-coating of
        iron-based metal pieces, Patent RO125457-B1/2014.

[15]   P.-E. Tegehal, N.-G. Vannerberg, Nucleation and formation of zinc phosphate
        conversion coating on cold-rolled steel, Corrosion Science, 32(5-6) (2012) 401-
        406. https://doi.org/10.1016/0010-938X(91)90112-3

[16]   C.Y. Tsai, J.S. Liu, P.L. Chen, C.S. Lin, A two-step roll coating
        phosphate/molybdate passivation treatment for hot-dip galvanized steel sheet,
        Corrosion Science, 52(10) (2010) 3385-3393.
        https://doi.org/10.1016/j.corsci.2010.06.020

[17]   F.S. Sayyedan, M.H. Enayati, Evaluating oxidation behavior of amorphous
        aluminum phosphate coating, Applied Surface Science, 455 (2018) 821-830.
        https://doi.org/10.1016/j.apsusc.2018.06.087

[18]   D.P. Burduhos-Nergis, A.V. Sandu, D.D. Burduhos-Nergis, D.C. Darabont, R.-I.
        Comaneci, C. Bejinariu, Shock Resistance Improvement of Carbon Steel

Carabiners Used at PPE, MATEC Web Conf. 290 (2019) 12004.
https://doi.org/10.1051/matecconf/201929012004

[19]   C. Nejneru, M.C. Perju, D.D. Burduhos Nergis, A.V. Sandu, C. Bejinariu,
       Galvanic Corrosion Behaviour of Phosphate Nodular Cast Iron in Different Types
       of Residual Waters and Couplings, REVISTA DE CHIMIE, 70(10) (2019) 3597-
       3602. https://doi.org/10.37358/RC.19.10.7604

[20]   D.P. Burduhos-Nergis, P. Vizureanu, A.V. Sandu, C. Bejinariu, Evaluation of the
       Corrosion Resistance of Phosphate Coatings Deposited on the Surface of the
       Carbon Steel Used for Carabiners Manufacturing, Applied Sciences 10(8) (2020)
       2753. https://doi.org/10.3390/app10082753

[21]   D.P. Burduhos-Nergis, P. Vizureanu, A.V. Sandu, C. Bejinariu, Phosphate Surface
       Treatment for Improving the Corrosion Resistance of the C45 Carbon Steel Used
       in Carabiners Manufacturing, Materials 13(15) (2020) 3410.
       https://doi.org/10.3390/ma13153410

[22]   A.V. Sandu, C. Bejinariu, G. Nemtoi, I.G. Sandu, P. Vizureanu, I. Ionita, C. Baciu,
       New anticorrosion layers obtained by chemical phosphatation, REVISTA DE
       CHIMIE, 64(8) (2013) 825-827.

[23]   P. Lazar, C. Bejinariu, A.V. Sandu, A.M. Cazac, O. Corbu, M.C. Perju, I.G.
       Sandu, Corrosion Evaluation of Some Phosphated Thin Layers on Reinforcing
       Steel, IOP Conference Series: Materials Science and Engineering, 209(1) (2017)
       012025. https://doi.org/10.1088/1757-899X/209/1/012025

[24]   D.P. Burduhos-Nergis, C. Nejneru, R. Cimpoesu, A.M. Cazac, C. Baciu, D.C.
       Darabont, C. Bejinariu, Analysis of Chemically Deposited Phosphate Layer on the
       Carabiners Steel Surface Used at Personal Protective Equipments, Quality-Access
       to Success 20(1) (2019) 77-82.

[25]   D.P. Burduhos Nergis, N. Cimpoesu, P. Vizureanu, C. Baciu, C. Bejinariu,
       Tribological characterization of phosphate conversion coating and rubber paint
       coating deposited on carbon steel carabiners surfaces, Materials today:
       proceedings 19 (2019) 969-978. https://doi.org/10.1016/j.matpr.2019.08.009

[26]   C. Bejinariu, P. Lazăr, A.V. Sandu, A.M. Cazac, I.G. Sandu, O. Corbu, Enhancing
       properties of reinforcing steel by chemical phosphatation, Applied Mechanics
       AND Materials, 754-755 (2015) 310-314.
       https://doi.org/10.4028/www.scientific.net/AMM.754-755.310

[27]   A.V. Sandu, C. Bejinariu, A. Predescu, I.G. Sandu, C. Baciu, I. Sandu, New

mechanisms for chemical phosphatation of iron objects, RECENT PATENT ON CORROSION SCIENCE, (ISSN 1877-6108), Bentham Science Publishers, 1(1) (2011) 33-37. https://doi.org/10.2174/2210683911101010033

[28]   D. Ernens, G. Langedijk, P. Smit, M.B. de Rooij, H.R. Pasaribu, D.J. Schipper, Characterization of the Adsorption Mechanism of Manganese Phosphate Conversion Coating Derived Tribofilms, Tribology Letters, 66 (2018) 131. https://doi.org/10.1007/s11249-018-1082-2

[29]   X.J. Yang, H.L. Jin, Phosphate and Stannate Chemical Conversion Coatings Formed on AZ91D Alloys, Materials Science Forum, 610-613 (2009) 1407-1013. https://doi.org/10.4028/www.scientific.net/MSF.610-613.1407

APPENDIX 1

Linear Polarization Curves in Semi-Logarithmic Coordinates

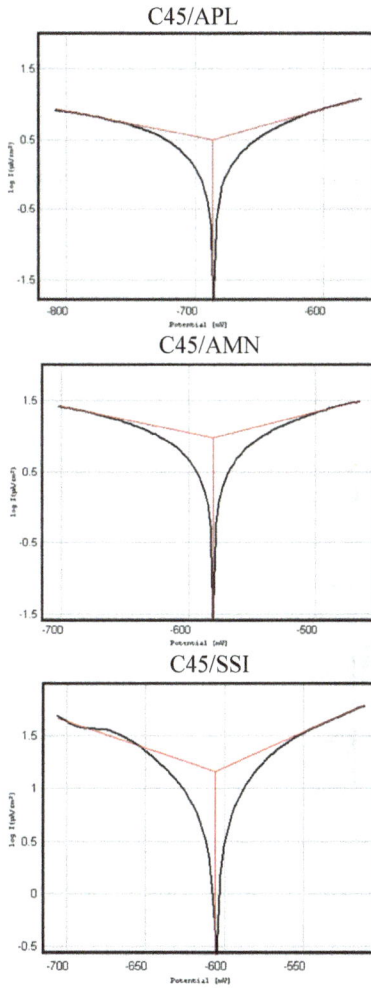

C45/APL

C45/AMN

C45/SSI

I-Zn/APL

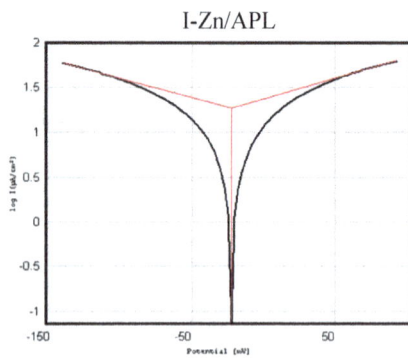

I-Zn/AMN

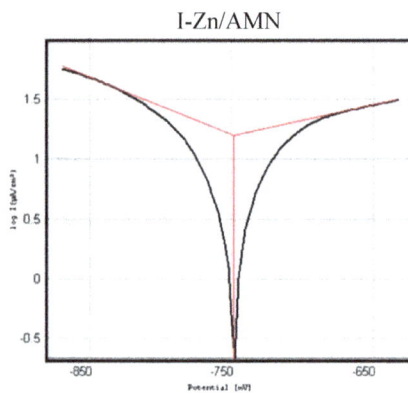

I-Zn/SSI

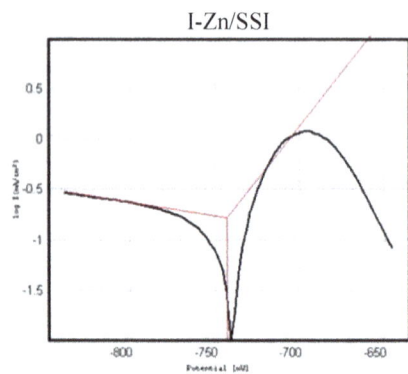

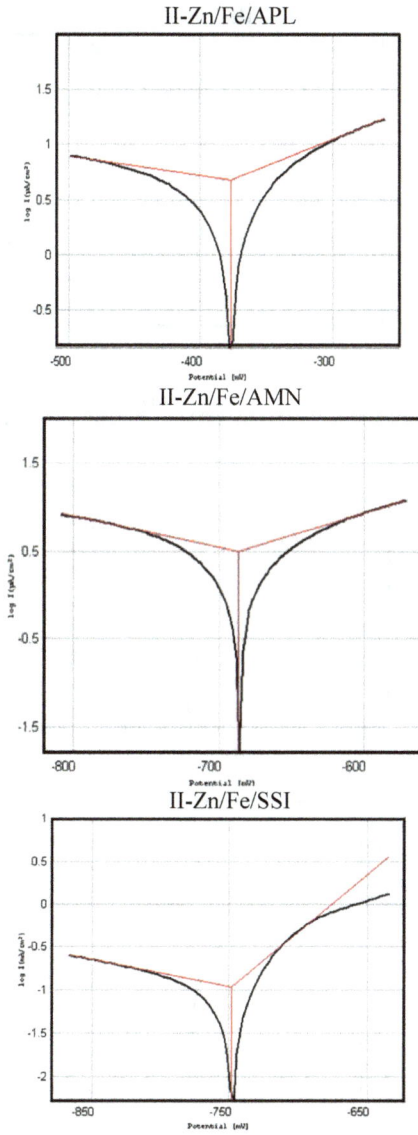

III-Mn/APL

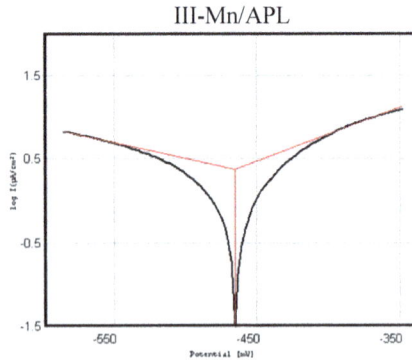

III-Mn/AMN

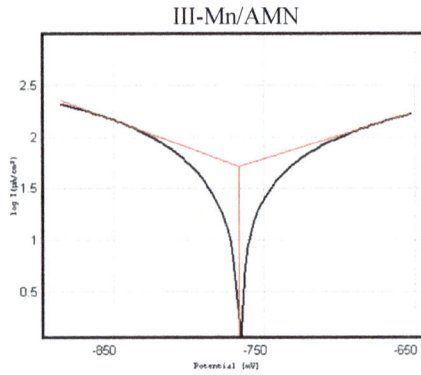

III-Mn/SSI

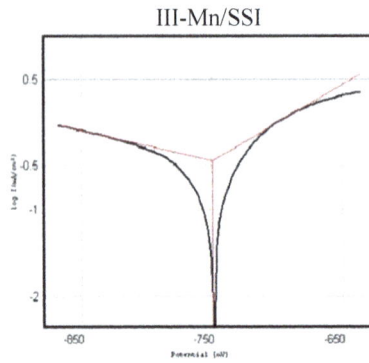

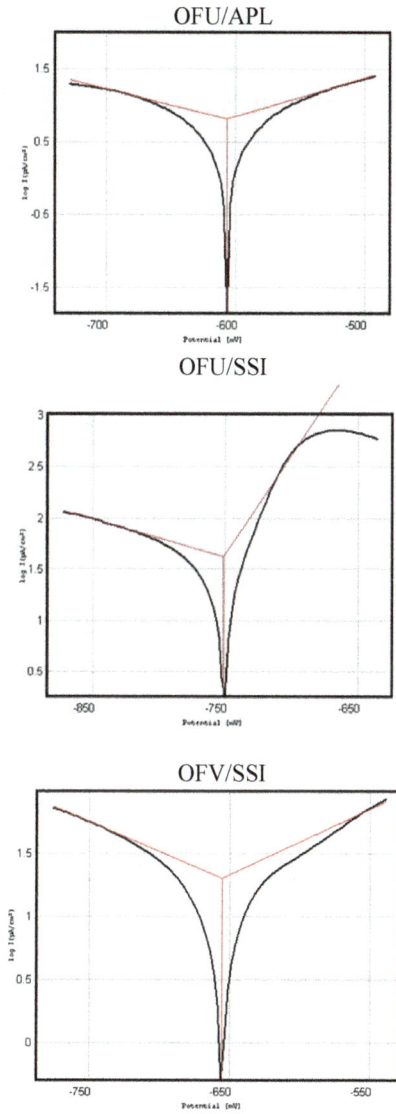

APPENDIX 2

Cyclic Voltamograms in Semilogarithmic Coordinates

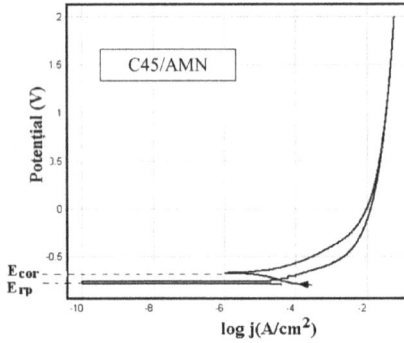

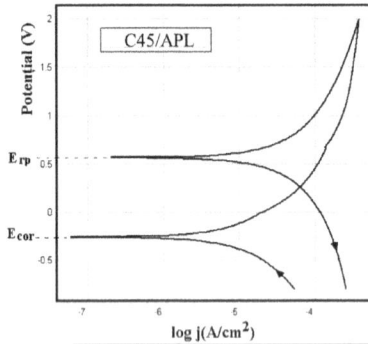

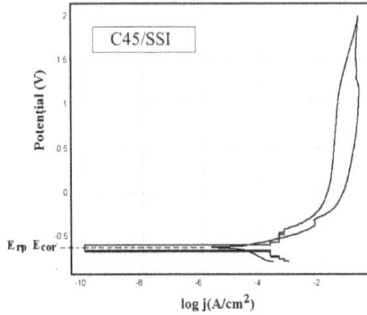

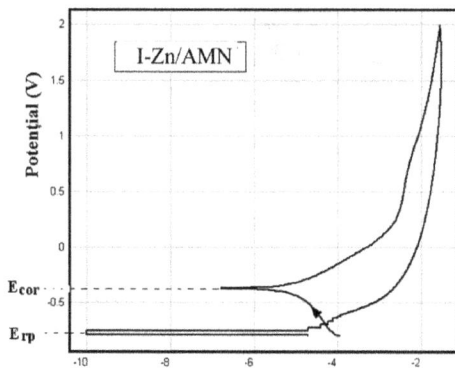

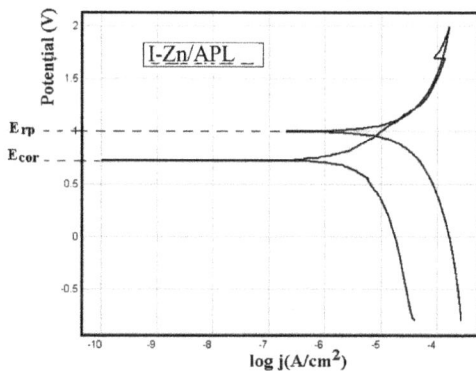

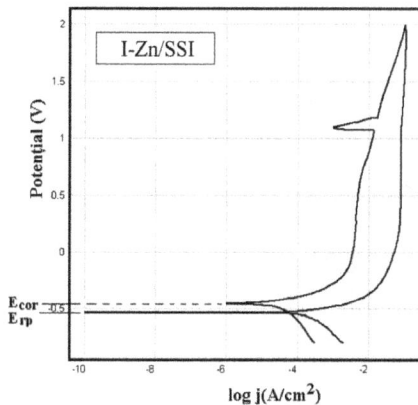

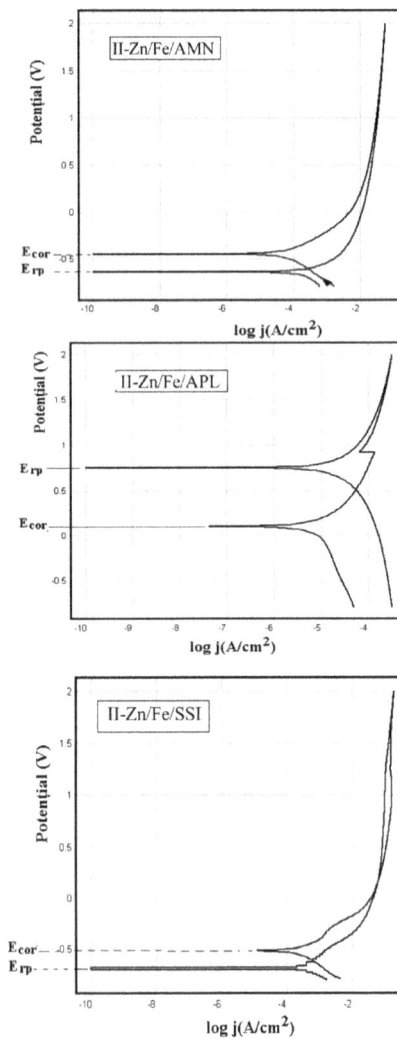

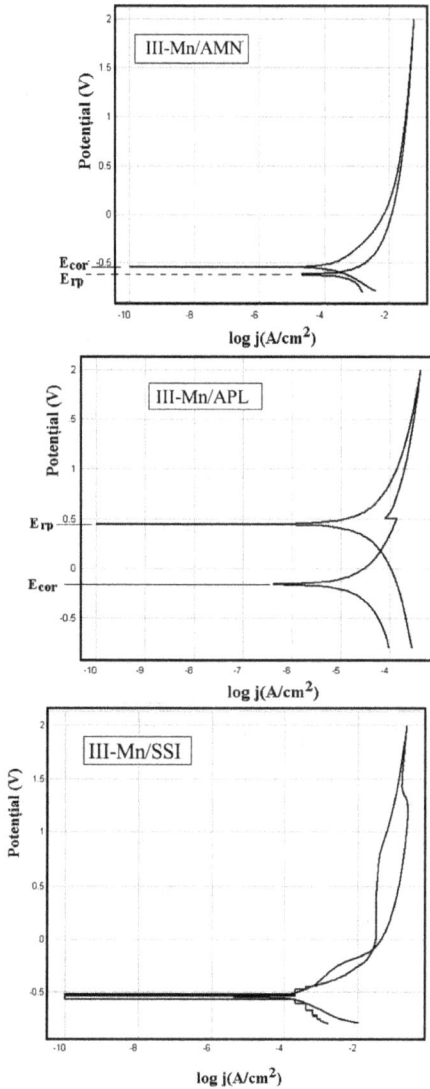

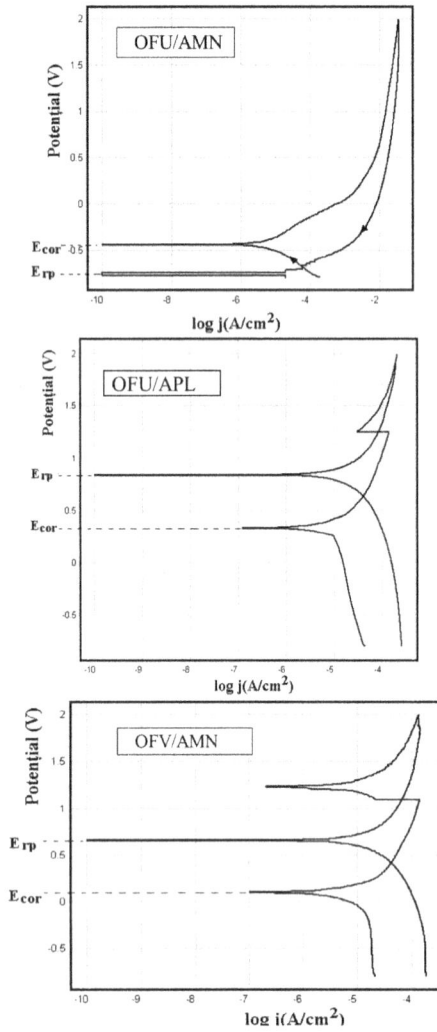

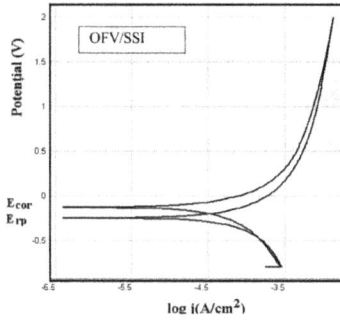

APPENDIX 3

Nyquist and Body Diagrams for Alloy/Corrosion Environment Systems

C45/AMN

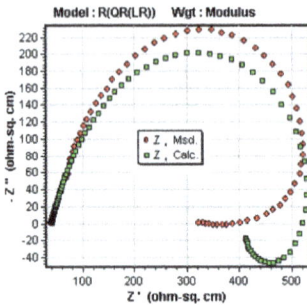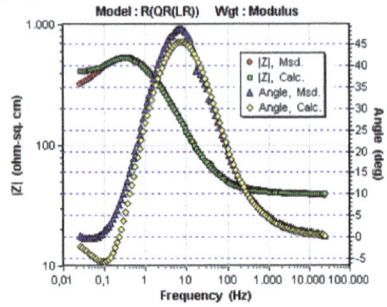

C45/APL

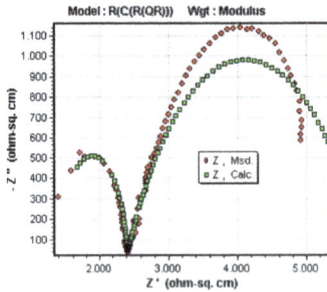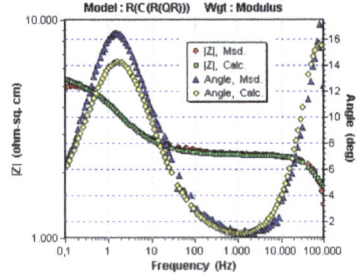

I-Zn/AMN

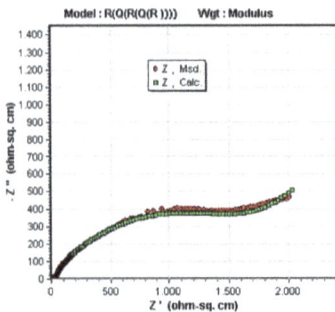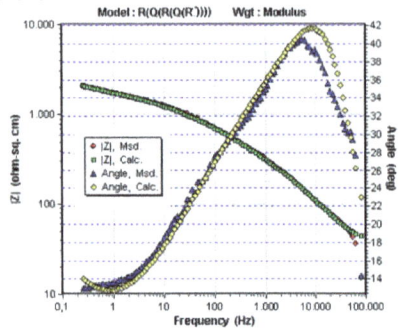

II-Zn/Fe/AMN

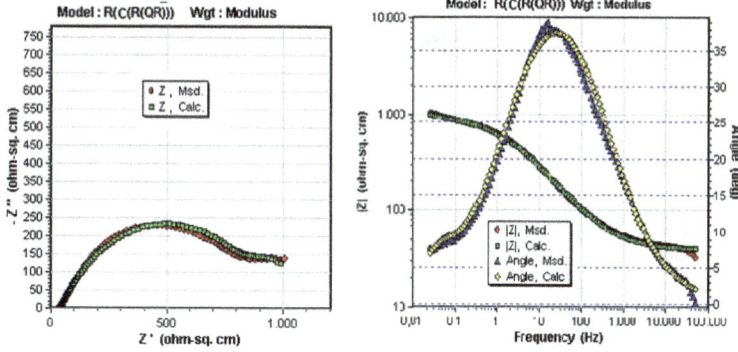

III-Mn/AMN

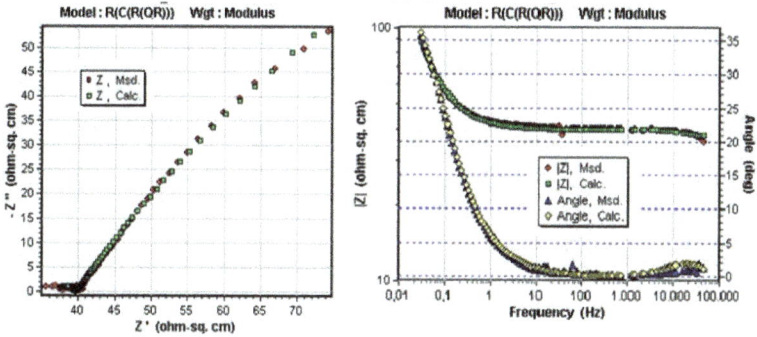

OFU/APL

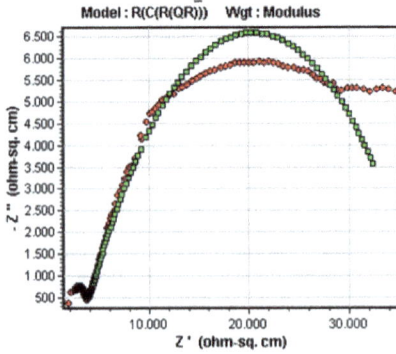
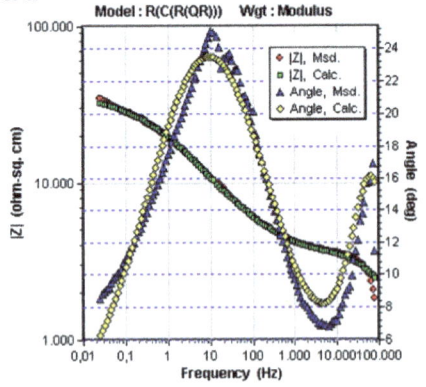

I-Zn/APL

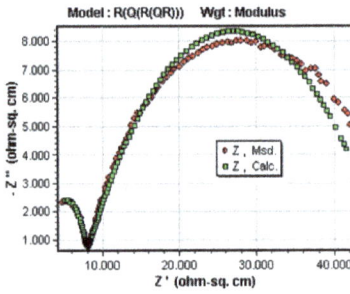
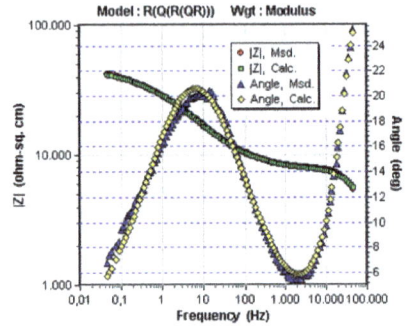

II-Zn/Fe/APL

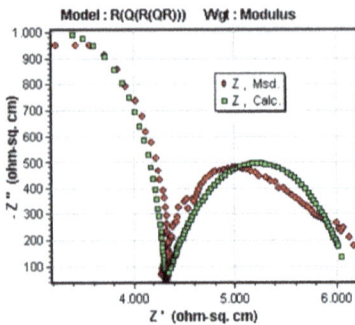
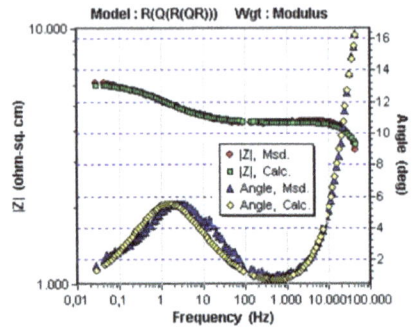

III-Mn/APL

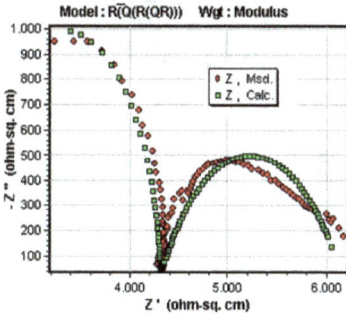
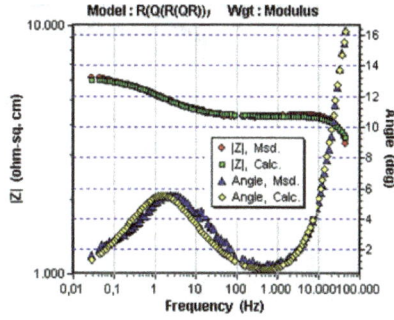

I-Zn/SSI

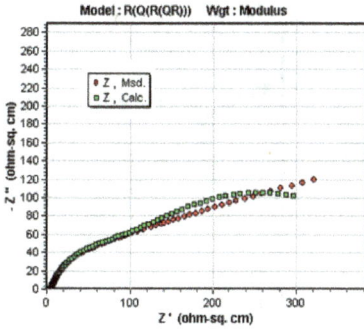
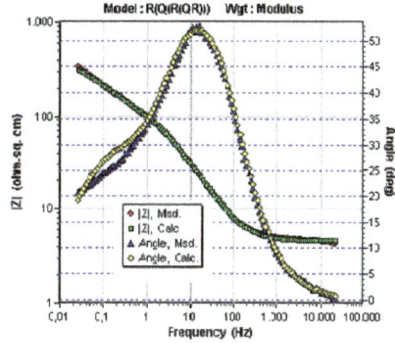

III-Mn/SSI

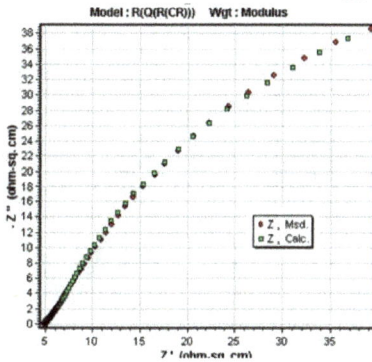
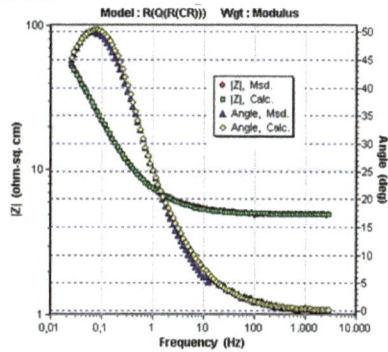

C45/SSI

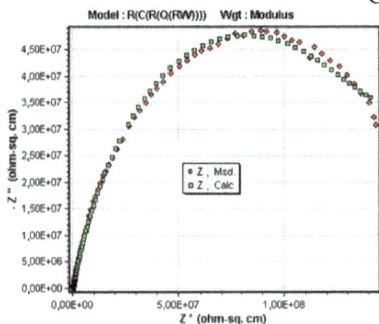
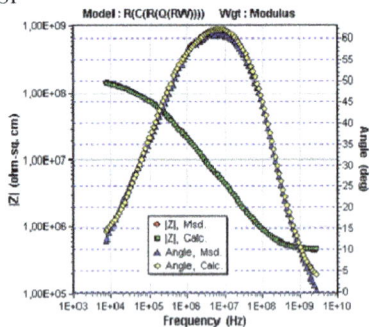

OFU/APL

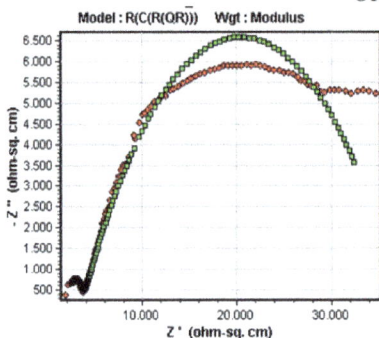
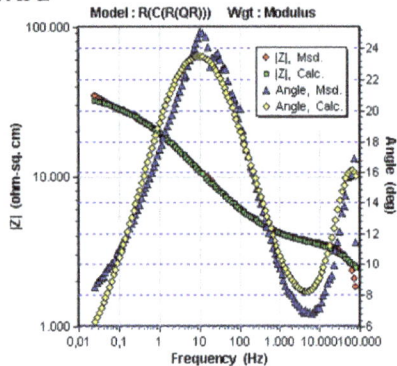

I-Zn/APL

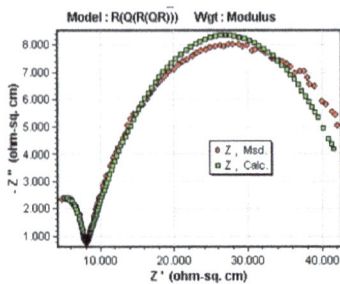
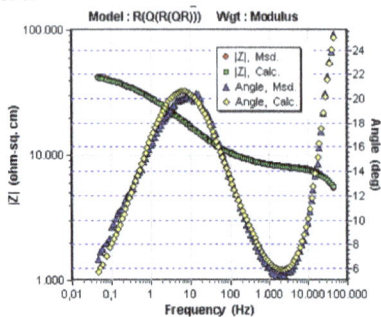

II-Zn/Fe/APL

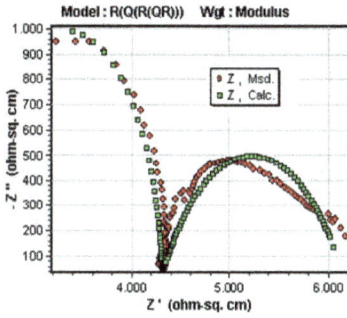
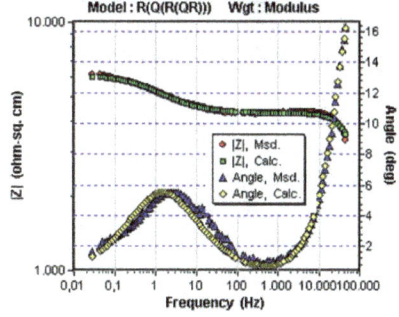

I-Zn/SSI

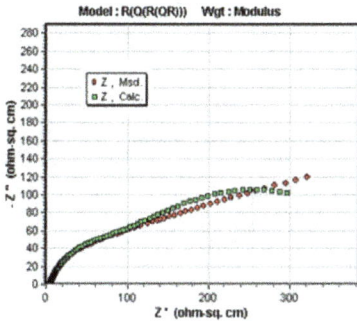
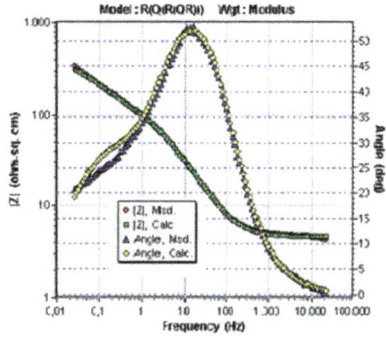

II-Zn/Fe/SSI

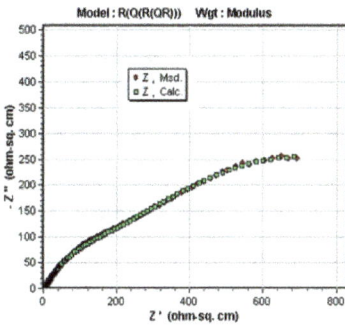
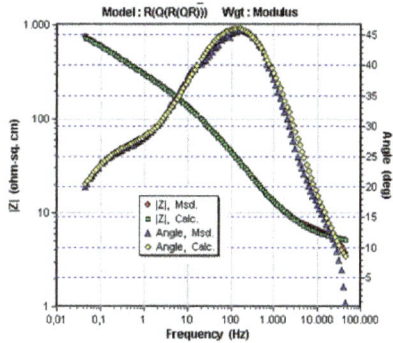

III-Mn/SSI

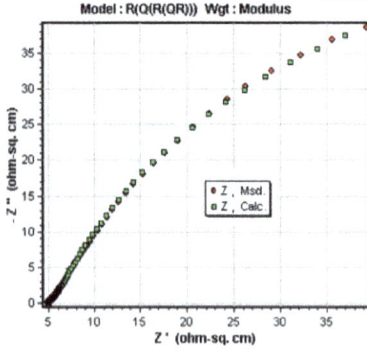

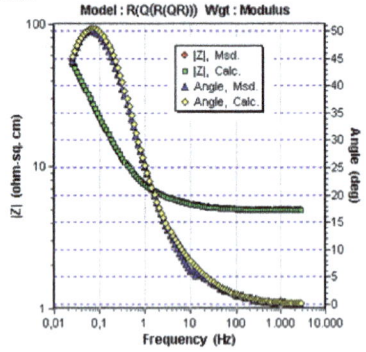

C45/SSI

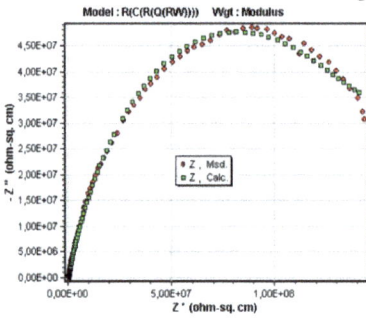

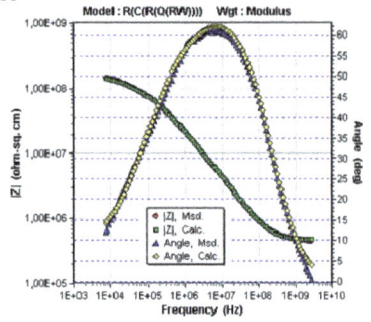

OFV/AMN

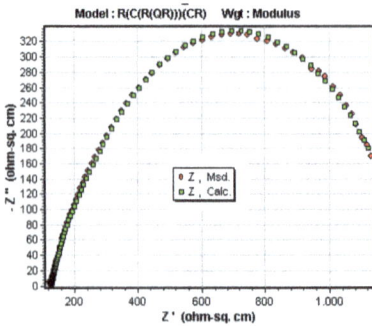

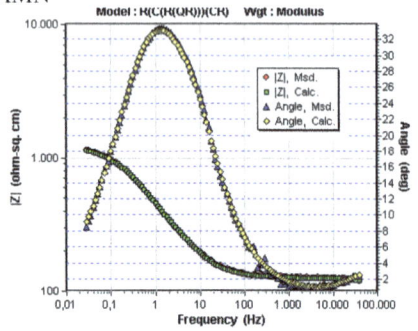

OFV/SSI

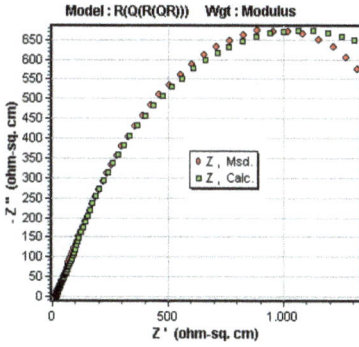

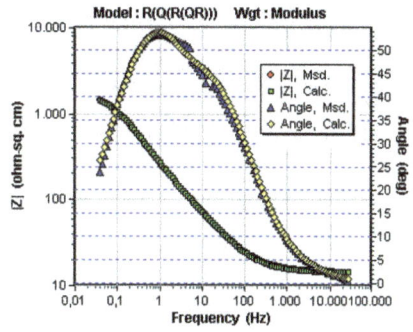

About the Authors

Diana Petronela BURDUHOS-NERGIS

Assistant Professor PhD.Eng.

Gheorghe Asachi Technical University of Iasi

diana.burduhos@tuiasi.ro, www.afir.org.ro/dpbn

Dr. Burduhos-Nergis is a researcher and assistant professor at the Gheorghe Asachi Technical University of Iasi, Faculty of Materials Science and Engineering, with a doctoral thesis on the study and improvement of carbon steel components in personal protective equipment. She is involved in scientific research since she was a student. She has over 15 publication, 14 of them indexed by Web of Science. She received many awards from presentations at conferences and invention exhibitions.

Costica BEJINARIU

Professor Ph.D. Eng.

Vicedean of Faculty of Materials Science and Engineering,

"Gheorghe Asachi" Technical University of Iasi

costica.bejinariu@tuiasi.ro

Dr. Bejinariu is a professor and researcher at the "Gheorghe Asachi" Technical University of Iasi, with more than 30 years of experience. He is a PhD Coordinator since 2009, with 5 granted PhD students and 10 undergoing PhD student. His field of experience is materials engineering with 18 published books and over 200 published articles which received more than 600 citations – H index of 14/WoS and 18/Scopus. He worked on more than 45 research grants, on 4 of these being director and another 3 as the institutions responsible. He has 12 patents and many awards received for these. He is a member of various academic societies and also reviewer for many scientific journals and conferences.

Andrei Victor SANDU

Associate Professor PhD.Eng. (Senior Lecturer)
Gheorghe Asachi Technical University of Iasi
President of Romanian Inventors Forum
sav@tuiasi.ro, www.afir.org.ro/sav

Dr. Sandu is a researcher and associate professor at the Gheorghe Asachi Technical University of Iasi, Faculty of Materials Science and Engineering. Dr. Sandu has expertise in the field of materials science, mainly on advanced analysis techniques. He has started his "scientific life" young with a first publication at the age of 18 years. Now he has over 300 publications, 250 of them indexed by Web of Science (Thomson Reuters) and over 30 patents. Regarding international recognition, the Hirsh index is 22 (over 1500 citations), being a visiting Professor at the Universiti Malaysia Perlis. On the innovative side, he has received over 100 medals at inventions exhibitions and contests and various important orders. He is the President of the Romanian Inventors Forum, member of WIIPA – World Invention Intellectual Property Associations and full member for Romania at IFIA – International Federation of Inventors' Associations. He is the publishing editor of the International Journal of Conservation Science (Web of Science and Scopus Indexed) and European Journal of Materials Science and Engineering and reviewer for many valuable journals.

www.ingramcontent.com/pod-product-compliance
Lightning Source LLC
Chambersburg PA
CBHW071227210326
41597CB00016B/1972